Noojin Walker
Martha Fulton Walker

The Twain Meet

The Physical Sciences and Poetry

PETER LANG
New York • Bern • Frankfurt am Main • Paris

Library of Congress Cataloging-in-Publication Data

Walker, Noojin
   The twain meet : the physical sciences and poetry / Noojin Walker and Martha Fulton Walker.
      p. cm. — (American university studies. Series XIX, General literature ; vol. 23)
    Bibliography: p.
    Includes index.
   1. Literature and science. 2. Literature and technology. 3. Poetics. I. Walker, Martha Fulton II. Title. III. Series.
PN55.W327   1989   809.1—dc20   89-12479
ISBN 0-8204-0953-7   CIP
ISSN 0743-6645

CIP-Titelaufnahme der Deutschen Bibliothek

Walker, Noojin:
The twain meet : the physical sciences and poetry / Noojin Walker; Martha Fulton Walker. — New York; Bern; Frankfurt am Main; Paris: Lang, 1989.
  (American University Studies: Ser. 19, General Literature; Vol. 23)
  ISBN 0-8204-0953-7

NE: Walker, Martha Fulton:; American University Studies / 19

© Peter Lang Publishing, Inc., New York 1989

All rights reserved.
Reprint or reproduction, even partially, in all forms such as microfilm, xerography, microfiche, microcard, offset strictly prohibited.

Printed by Weihert-Druck GmbH, Darmstadt, West Germany

The Twain Meet

American University Studies

Series XIX
General Literature

Vol. 23

PETER LANG
New York • Bern • Frankfurt am Main • Paris

Acknowledgements

"First and Second Law" by Michael Flanders from *Faber Book of Useful Verse* copyright 1981. Permission granted by Claudia Flanders, executor of Michael Flanders Estate.
"Curb Science?" by Robinson Jeffers. Reprinted with the permission of the Estate of Robinson Jeffers, Lee H. Jeffers (Mrs. Donnan Jeffers for Jeffers Literary Properties)
"Oxygen" by Joan Swift. Reprinted by permission; copyright 1962 The New Yorker Magazine Inc.
"Watching Television" by Robert Bly from *Lights Around the Body* copyright 1962.
"Happiest Heart" by John Vance Cheney from *Poems* published by Houghton Mifflin Company
"Cloud, Horizon, Water" by Thomas Ockerse from *Speaking Pictures* edited by Milton Klonsky copyright 1975. Permission granted by Thomas Ockerse
"On Nature" from *A Source Book in Chinese Philosophy* translated by Wing'Tsit Chan. Permission to reprint granted by Princeton University Press
Selections from *Lucretius: The Way Things Are* translated by Rolfe Humphries. Copyright 1968. Reprinted by permission of Indiana University Press.
"Song of the Road" and "Ice Cream Cone" by Robert Gross from *Open Poetry* by Robert Gross and George Quasha. Copyright 1973. Permission to reprint by the publisher Simon and Schuster, Inc.
"Einstein" from *New and Collected Poems* 1917-1982 by Archibald MacLeish. Copyright (c) 1985 by The Estate of Archibald MacLeish. Reprinted by permission of Houghton, Mifflin Company.
Excerpts from *The People, Yes* by Carl Sandburg, copyright 1936 by Harcourt Brace Jovanovich, Inc. renewed by Carl Sandburg, reprinted by permission of the publisher.
"The Hollow Men" in *Collected Poems* 1909-1962 by T.S. Eliot, copyright 1963 by Harcourt Brace Jovanovich, Inc. copyright (c) 1963, 1964 by T.S. Eliot. Reprinted by permission of the publisher.
"For Whitman" copyright (c) 1973 by Diane Wakoski. Reprinted with the permission of the author.
"As I Walked Out One Evening" copyright 1940 and renewed 1968 by W.H. Auden. Reprinted from *W. H. Auden: Collected Poems,* Edited by Edward Mendelson, by permission of Random House, Inc. Originally appeared in *The New Yorker.*
"The Glass of Water" copyright 1954 by Wallace Stevens. Reprinted from *The Collected Poems of Wallace Stevens,* by permission of Alfred A. Knopf, Inc.
"The Great Scarf of Birds" copyright (c) 1962 by John Updike. Reprinted from *Telephone Poles and Other Poems* by John Updike, by permission of Alfred A. Knopf, Inc. Originally appeared in *The New Yorker.*
"Science" copyright 1925 and renewed 1953 by Robinson Jeffers. Reprinted from *Selected Poetry of Robinson Jeffers* by permission of Random House, Inc.
"Accidentally on Purpose", "Fire and Ice", "Innate Helium", "For Once, Then, Something", "Nothing Gold Can Stay". From *The Poetry of Robery Frost* edited by Edward Connery Lathem. Copyright (c) 1969 by Holt Rinehart and Winston, Inc. Copyright (c) 1962 by Robery Frost. Copyright (c) 1975 by Lesley Frost Ballanitne. Reprinted by permission of Henry Holt and Company, Inc.
Poems numbered 258,812,129 by Emily Dickinson. Reprinted by permission of the

publishers and the Trustees of Amherst College from *The Poems of Emily Dickinson,* edited by Thomas H. Johnson, Cambridge Mass.: The Belknap Press of Harvard University Press, Copyright 1951 (c) 1955, 1979, 1983 by the President and Fellows of Harvard College. "B:" (c) by the President and Fellows of Harvard College.; (c) renewed 1977 by Mrs. Eloise Bender.

"How Everything Happens (Based on a Theory of Waves)" by May Swenson is used by permission of the author, Copyright (c) 1969 by May Swenson.

"The Universe" by May Swenson is used by permission of the author, Copyright (c) 1963 by May Swenson.

"3 Models of the Universe" by May Swenson is used by permission of the author, Copyright (c) 1963 by May Swenson.

Poem numbered 341 by Emily Dickinson from *The Complete Poems of Emily Dickinson,* edited by Thomas H. Johnson. Copyright 1929 by Martha Dickinson Bianchi; copyright (c) renewed 1963 by Mary L. Hampton. By permission of Little, Brown and Company.

"E=MC$_2$" by Morris Bishop. From *The Best of Bishop: Light Verse from The New Yorker and Elsewhere* (Cornell). (c) 1946, 1974 Alison Kingsbury Bishop. Originally in *The New Yorker.*

"Warty Bliggens the Toad" from *Archy and Mehitabel* by Don Marquis. Copyright 1927 by Doubleday, a division of Bantam, Doubleday, Dell Publishing Group, Inc. Reprinted by permission of the publisher.

"In Time Like Air" is reprinted from *Collected Poems (1930-1973)* by May Sarton, by permission of W.W. Norton & Company, Inc. Copyright (c) 1974 by May Sarton.

"Reactionary Essay on Applied Science" from *Times Three* by Phyllis McGinley. Copyright 1951 by Phyllis McGinley. All rights reserved. Reprinted by permission of Viking Penguin, a division of Penguin Books USA, Inc.

Table of Contents

	Preface	1
I	Introduction	5
II	The Atom	17
III	Order	39
IV	Chemistry	67
V	Light	103
VI	Sound	121
VII	Earth and Sky	143
VIII	Matter	181
IX	Relativity and Truth	219
X	Acclaim and Disclaim	245
	References	289
	Index	297

PREFACE

"East is east and west is west / And never the twain shall meet." Rudyard Kipling's logic is valid -- they that go in opposite directions should not be expected to meet. Thus, it is from this same basis of logic that many of us perceive the physical sciences and poetry -- never the twain shall meet.

We know from experience, however, that east and west do meet. When two travelers set out in opposite directions, the immediate perception is that they move farther and farther apart. As we continue to observe them from our removed vantage place, however, we discover that in reality they are drawing closer to each other, until at last they meet on the other side. Is it true, then, that other travelers in what seem to be opposite directions -- physical sciences and poetry -- shall also meet? Physical sciences and poetry: Do the twain meet?

Science and poetry would seem to have very little in common. Science, stereotypically, is thought of as being precise and concrete -- poetry as being ambiguous and abstract. The scientist is thought of as being cold and calculating -- the poet as being warm and romantic. Consider what the British scientist and author, C. P. Snow (1905-1980) wrote in his essay entitled "The Two Cultures."

> The intellectual life of the whole of western society is increasingly being split into two polar groups. Literary intellectuals at the one pole ... at the other ... the physical scientist ... between the two a gulf of mutual incomprehension ... but most of all a lack of understanding.[1]

Snow wrote of the two separate worlds of the sciences and the arts -- or rather, the science world and the non-science world. He was concerned and distressed by this "gulf of mutual incomprehension."

Is our intellectual life analogous to the concept of the continental drift whereby a once unified whole has fractured and is drifting apart, leaving only discrete islands of knowledge separated by vast oceans of ignorance? Snow envisioned this, and he was fearful that we might be tempted to sit back "complacent in our unique tragedy" and to accept the fearful consequences of this "widening gap of ignorance."

And yet one should temper Snow's pessimism with the view that was expressed by the scholar and priest, John Henry Cardinal Newman (1801-1890), in his book *The Idea of a University*.

> Summing up, gentlemen, what I have said, I lay it down that all knowledge forms a whole, because its subject matter is one. For the universe in its length and breadth is so ultimately knit together, that we cannot separate off portion from portion ... They all belong to one and the same circle of objects; they are one and all connected together.[2]

If "all knowledge forms a whole," then is it our arbitrary definitions that cause fractures, separations, and exclusions? Have we, for various reasons, created separations where none should truly exist? Is a separation between science and art real or contrived? Certainly differences exist; and one cannot argue that they do not. But, the questions relate to the extent of the differences. This book looks at science and at poetry in terms of their oneness; not excluding their differences, but with the hypothesis that a symbiotic relationship between science and poetry has existed and shall continue to exist amidst the concerns and disagreements.

Ours is an era of science, a new era of science if you please, but every era has been the era of poetry. The ideas of evolution, the expanding universe, the atom, relativity, and the chemistry of living organisms were all beyond the imagination of man just a few generations ago. And yet, as great accomplishments of man, they cannot be excluded from having an impact on, and a relationship with, literature, and especially poetry.

As we explore our era of science and poetry, we shall examine how the two forms of expression of each are interwoven, but similarly, we shall examine the causes for the tendencies to separate them. To do this we shall present a brief summary of the historical development of science, the poetic attitudes about science and the philosophical-religious thought during the past four hundred years. After considering the evidence of the basic differences between science and poetry, we shall nonetheless document the proposition that science and poetry are inseparable expressions of the creative human spirit. By concrete examples we shall experience how poets have proclaimed and condemned the accomplishments of science, and how poets have used the concepts of science to enrich their works and to give them special meanings beyond the exact words.

CHAPTER I INTRODUCTION

> East is east and west is west,
> And never the twain shall meet.
> > -from "Ballad of the East and West"-
> > -Rudyard Kipling

"East is east and west is west / And never the twain shall meet." Rudyard Kipling's logic is valid -- they who set out in opposite directions should not be expected to meet. Thus, it is from this same basis of logic that many of us percieve the physical sciences and poetry -- never the twain shall meet. But is it true?

Historical overview

Mysticism, myth, and the geocentric universe were the sources of explanation of natural phenomena during the sixteen centuries following biblical times. But events of the late sixteenth century and the seventeenth century ushered in a new philosophy -- that of experimental science.

Although one finds it difficult to pinpoint an exact moment in history when a pattern of scientific thought emerges, we could well define a beginning in 1543. For fifteen hundred years the geocentric idea of Ptolemy's second century theory of the universe had been essentially unchallenged. His proposition that the earth was the center of the universe and that all celestial bodies revolved around it was not only accepted among the scholars of the times but also was dogma of the Church. But in 1543, Nicolaus Copernicus challenged the Ptolemic theory with his proof that the planetary motion did not support the idea that the earth was the center of the universe. The motions of the heavenly bodies could be explained only if the earth itself was moving through space and orbiting the sun.

As might be expected, however, the end to Ptolemy's theory did not come quickly. Almost a hundred years were to pass before Galileo's publication of the Copernican theory, and his subsequent trial by the Inquisition. Meanwhile, in 1609, Johanne Kepler introduced the laws of planetary orbits. He described a pattern of orderliness within the heavens that, although contrary to the idea that the earth was the center of the universe, reinforced the prevailing religious belief in God's masterful creation.

During the same period two other inventions gave to man, literally, a new view of his universe -- the microscope, in 1590, and the telescope, in 1608. The new poetry dramatized the new inventions, and the theories and discoveries that resulted. The philosophical concept of the great Chain of Being received new emphases in as much as the orderly life in a drop of stagnant pond water was interpreted as more evidence of God's magnificence. And as telescopic observations of distant stars and of the orderly celestial motion increased, the idea of the possibility of inhabitants on "other worlds" in the universe increased. John Dryden (1631-1700), poet laureate of England, expressed this in his poem "Elenora."

Perhaps a thousand other worlds that lie

Remote from us, and latent in the sky,
Are lighted by His beams, and kindly nursed.
       - from "Elenora"-
       - John Dryden-

Thus scientific thought that could have been judged blasphemous a short time before now became accepted proof of an even more omnipotent Creator. Yet, others saw among the falling stars and glowing comets perhaps an omen of the end of the world. William Drummond's "The Shadow of Judgement," written in 1630, was a case in point.

New worlds seen, shine
With other suns and moons, false stars decline,
And blaze, as other worlds are judged there.
       -from "The Shadow of Judgement"-
       -William Drummond-

Still, the same scientific discoveries caused other poets, such as Alexander Pope (1687-1704), to turn even more inward expressing, in effect, that one should be concerned only with oneself. The point was made clearly in a passage of his poem, "Essay on Man", Epistle I:

Thro' worlds unnumber'd tho' the God be known,
'T is ours to trace him only in our own.

William Shakespeare (1564-1616) was contented with the medieval view of astronomy and Christopher Marlowe (1564-1593) continued to regard the works of Aristotle as the final word in science. But irreversible changes were heralded by Issac Newton's *Principia* (1687) and *Optics* (1704). *Principia* introduced his discoveries of the laws of motion and the theory of gravitation. As with all great ideas, each could be simply and powerfully described. The force he defined as universal gravitation resulted from the attraction each body has upon another. And the magnitude of the force was dependent upon the amount of matter in each and the distance between them.

In 1704, the results of Newton's experiments with light were published in *Opticks*. He showed that the "pure white light" was neither pure nor white, that it was composed of a series of colors from violet to yellow to red and that each color could be separated from the others. Newton wrote: "The most surprising and wonderful composition was that of 'whiteness'. Whiteness is the usual color of Light; for Light is a confused aggregate of Rays induced with all sorts of colors."

Newton's new sense of color seemed to spread throughout western civilization just as the spectrum had spread its light across his wall. Jacob Bronowski (1908-1974), mathematician, scientist and author, writes that "the palette of painters became more varied, . . . and that it became natural to use many colour words." "Pope," he said, "was surely a less sensuous poet than Shakespeare, yet he uses three or four times as many colour words as Shakespeare, and uses them about ten times as often." For example, consider Pope's description of fish in the River Thames:

> The bright-ey'd Perch with Fins of Tyrian Dye,
> The silver Eel, in shining Volumes roll'd,
> The yellow Carp, in Scales bedrop'd with Gold,
> Swift Trouts, diversify'd with Crimson Stains.[1]

Newton's discovery of the cause of the spectrum and the force that he labeled universal gravitation, together with his creation of the calculus, dawned a period of enthusiastic acceptance and praise of the advancement of science by quasi-progressive poets such as James Thompson, Henry Brooke, and Mark Akenside.

The nineteenth century opened with John Dalton's formulation of the atomic theory in 1808, followed by Charles Darwin's *Origin of the Species* in 1859. These theories exacerbated a hostility that had begun to develop between certain influential poets and the new science. Although Alexander Pope,

Christopher Smart (1722-1771), and William Cowper (1731-1800) used the discoveries of science to glorify God, they realized that the arrogance of science was increasing; it refused to admit to its own limitations. Blake rejected science even more because to him it was the enemy of imagination. Consequently a school of poets was emerging that opposed the materialistic analysis and experimentation of science as being in conflict with the spiritual world of imagination.

Toward the latter part of the nineteenth century, two main reasons led to the disillusionment of certain poets with science and the scientist. The romantic poets held firm to their belief that science was leading to a mechanical view of the universe. This was combined with an increasing repugnance to technology and the industrialization of society. Pollution, congestion, and exploitation reinforced their hostility and made them harken back to a pre-science simplicity and beauty as expressed by William Morris (1834-1896).

> Forget six counties overhung with smoke,
> Forget the snorting steam and piston stroke,
> Forget the spreading of the hideous town;
> Think rather of the pack-horse on the down,
> And dream of London, small, and white, and clean,
> The clear Thames bordered by its garden green.
> - from "Prologue to the Earthly Paradise"-
> - William Morris-

As the twentieth century sweeps before us, it is safe to say that no poet today, to whichever school he belongs, can ignore science. For interwoven within the scientific upsurge -- Einstein's Relativity in 1905, Heisenberg's Uncertainty in 1927, Fermi's atomic age in 1942, and Sputnik in 1957 -- is the whole political/social movement of the last three centuries, and most especially of the last hundred years. Nonetheless, some modern poets continue to see a conflict existing between themselves and science because of their perception of a basic incompatibility between the visionary powers of the poets and the analytical methodology of the scientist. Others believe the conflict to be between those who romanticize the past as opposed to those who accept the future. One group of poets views twentieth century man as frightened, anxious, incompetent, a victim of science and his own possible self-destruction. The other group sees man as the beneficiary of science -- uplifted and strengthened.

The uncommon common ground

The essence of science is that it deals with the familiar elements of our experience. Science is concerned with rocks, leaves, stars, heat, sound, electricity, life, change, movement, and truth. Every rock is a snapshot of history; every action creates an equal and opposite reaction; matter and energy are constant. The scientist studies these elements. He exposes their regularities, their predictabilities, their patterns, and their anomalies, and amorally shapes them to man's use. From the raw materials of the physical world he brings into existence new discoveries. Professor Hymen Levy gives us this interesting perspective:

> Every discovery is an explanation. Every explanation is a discovery. He is explaining the world. He is exposing newness as he does. In this lies the creativeness of the scientist and it is in very much the same way that the poet shows creativeness.[2]

The essence of poetry is that it too deals with the familiar elements of our experience. Poetry is concerned with love, hope, being, continuance, self, suffering, diety, and truth. The poet reflects the immediate conflicts, uncertainties, and problems of his day, but not mechanically as an image is reflected from a mirror. In the broadest sense he is using his emotional and intellectual understanding to analyze, and to reorganize the stimuli into a new synthesis. From the raw materials of experience he fashions something new that others may share.

> The great power of poetry is the interpretative power; by which I mean not a power of drawing out in black and white an explanation of the universe, but the power of so dealing with things as to awaken in us a wonderfully full, new, and intimate sense of them. When this sense is awakened in us, as to objects without us, we feel ourselves to be in contact with the essential nature of these objects, to be no longer bewildered and oppressed by them.[3]
> -Matthew Arnold (1822-1888)-

Thus we find the similarity in purpose of science and of poetry. Each of them searches, discovers, and communicates. Each approaches the fact/value problem of the era of science. Professor Hyatt Howe Waggoner, author and critic, summarizes the problem best when he describes it as, "The place of fact in a world of value; the place of value in a world of fact."[4] Each engages the same basic disquiet -- man's longing for reasons, and ultimately a Reason.

Although one may intuitively question the idea that poets and scientists have any common ground between them, or that any of the same tools of the trade are used by both, one common denominator clearly surfaces -- words. And words, we are told, have two functions: to state facts and to evoke emotions. If the first function belongs to science and the second to poetry, is it dangerous to mix them?

One should not deny the distinction and the tension between science and poetry. But the facts as reported by science need not inhibit the poet's mission. The poet succeeds in expressing the deepest experiences while the scientist succeeds, in touching only the most superficial ones. The scientist has the slight compensation, however, that his conclusions can be verified by anyone. For the scientist the words are precise; words are selected that will eliminate all but one meaning. We believe the scientist because he can substantiate his remarks, not because he is forceful or eloquent in his presentation. In fact, we distrust him somewhat when he seems to be overly persuasive in his manner.

In most poetry, the use of words is almost the reverse of science. The poet makes the reader pick out the precise meaning from an indefinite number of possible meanings which a word, a phrase, or a sentence may convey, As the British critic and semanticist, I. A. Richards, writes, "It is never what a poem *says* which matters, but what it is."[5] But how can the words be divorced from the meaning?

> A good deal of poetry and even some great poetry exists in which the sense of the words can be almost entirely missed or neglected without loss. ... the sound and the feel of the words, what is often called the form of the poem in opposition to its content, get to work first, and the senses in which the words are later more explicitly taken are subtly influenced by the fact. Most words are ambiguous as regards their plain sense, especially in poetry ... The senses we are pleased to choose are those which most suit the impulses already stirring and giving form to the verse. Thus the form often seems as an inexplicable premonition of a meaning which we have not yet grasped. The same thing can be noticed in conversation. Not the strict logical sense of what is said, but the tone of voice and the occasion are the primary factors by which we interpret.[6]
> -I.A. Richards-

The poet in his quest for Reason, relies in part upon logic and in part upon imagination. Few would seriously challenge the premise that science and mathematics have contributed significantly to the development of logical thought and systematic processes. The belief of many, however, has been that science is the Nemesis to imagination. It has been this influence on imagination which has been the most disturbing to the poets. John Keats (1795-1821) firmly believed that science was hostile to imagination for why else would Newton destroy beauty by explaining the rainbow as a prismatic spectrum. And William Blake (1757-1827), the poet and mystic, could never forgive scientists for having analyzed the divine mysteries of natural experiences into little more than physical measurements. And replacing the elements of air, earth, fire, and water with inanimate spheres of identical atoms, and describing the rainbow as little more than bent light which anyone could produce by holding a glass prism in the path of the rays, was anathema to someone like Wordsworth. It was his fertile imagination which had moved him to exclaim so enthusiastically, "My heart leaps up when I behold / A Rainbow in the sky!"

> Thus for the scientist "water becomes H_2O; a pond is H_2O, with an index of refraction and a bottom composed of precisely named minerals with a certain weight per specific volume. (But the poet) knows that water in the form of rain is different from that in the form of a flood; and no knowledge derived simply from a scientific description will explain the realities of quenching thirst or destruction by flood."[7]
> -Norman Holmes Pearson-

The mystic is satisfied with the natural experience itself; the artist and poet attempt to express the experience as a connected system. In this sense both the poet and the scientist employ imagination even though for different reasons and by different means. The poet uses imagination for the purpose of recreating in others the experience that he has lived or felt. In this way his imagination is freer than that of the scientist; it is not required to stay within boundaries; it does not have to conform to patterns that have been established or that are yet to be established.

The scientist, on the other hand, must use his imagination within a single context as he develops a continuation of patterns found between one experience and another. Great scientist have been men with fertile imaginations. Creativity that involves going beyond the known yet staying within the boundaries set by the known, the inspiration, and the intuitive leaps, are processes of these imaginative minds. What other mental processes could have been at work when the chemist, Kekule (1829-1896), plagued by the inconsistencies of the accepted structure of benzene dreamed of snakes?

> I was busy writing on my textbook but could make no progress -- my mind was on other things. I turned my chair to the fire and sank into a doze. Again the atoms were before my eyes. Little groups kept modestly in the background. My mind's eye, trained by the observation of similar forms, could now distinguish more complex structures of various kinds. Long chains here and there more firmly joined; all winding and turning with snake-like motion. Suddenly one of the serpents caught its own tail and the ring thus formed whirled exasperatingly before my eyes. I woke as by lightning, and spent the night working out the logical consequences of the hypothesis. If we learn to dream we shall perhaps discover truth. But let us beware of publishing our dreams until they have been tested by the waking consciousness.[8]
>
> - Friedrich August Kekule -

The last thought expressed by Kekule in the quotation above has been expressed by H.G. Wood as a difference between the scientist and the poet.

The scientist must not be deflected from his own proper task by the pleasures of the poetic imagination. Science must exclude all wishful thinking, if it is to satisfy the one desire to know the truth, whatever it be. Poetry is shot through with what seems to be wishful thinking since it gives expression to all sorts of desires, hopes, fears, loves and hates.[9]

- H. G. Wood -

Yet, if there is one characteristic that is seen as setting the scientist apart from the poet, and indeed from the other artists, it is the experimental approach. The age of science is the age of experimentation. Francis Bacon (1561-1626), philosopher and writer of the seventeenth century, saw that imagination and unverified belief was decreasing as investigation and experimentation increased. Even as he announced the ascendancy of experimentation, he could not have realized all of the consequences.

But just as imagination is not an exclusive tool of the poet, neither is experimentation reserved only for the scientist. The poet attempts different words, varies the pattern, changes the sound, modifies the form -- all in an experimental process to recreate within the reader/listener the poet's experience. This is experimentation, albeit experimentation with a difference. The scientist's experimentation is a search for truth, in which case truth is defined as replicability. He accomplishes his goal when the truth he discovers can be replicated by other scientists. He utilizes a process for creating unvarying results.

The poet, however, does not create a pattern of constancy and replicability. His goal is not a formula. And whereas the work of the scientist is accumulative within the continuum of scientific knowledge, the poet must begin anew. The work he used before cannot be used again merely because it worked in the past. The mood created before cannot be transferred in toto to his new effort. The scientist builds upon the past; the poet does not so easily inherit the past.

CHAPTER II   THE ATOM

2. Poetry is the journal of a sea animal living on land, wanting to fly the air.
>   - from "Ten Definitions of Poetry" -
>   - Carl Sandburg -

Natural Philosophy

Man early observed that in his physical world everything changes, yet nothing changes. The rivers flow into the sea, but are not emptied. The sun sets, but rises again tomorrow. Plants die, but the world remains covered in vegetation. It was within this paradox that the ancient Greeks contemplated the structure of matter. The paradox serves us well today to show why philosophy was the monolithic discipline of inquiry among the ancient scholars; science per se did not exist nor was there a need for it to exist. The natural world was explained through natural philosophy.

Science requires data, and for the most of the ancient, medieval, and Renaissance periods, the available data were little more than common experiences. As such, natural phenomena were easily accommodated within the methodology -- the methodology of philosophy. In their quest for the explanation of the basic component of the material world, the ancient Greeks had reduced the explanation to the "permanence." The basic common component to all matter was that which was the most permanent. Thales of Miletus, about the seventh century B.C., proposed that water, because of its apparent permanence, was the ultimate substance -- it was primary matter.

In the mid 500s B.C., Anaximenes reasoned that air was the common unity. Later, Heraclitus altered the logic and defined change, not permanence, as the fundamental component. And because fire was ever-changing, it obviously was the basic element of all matter. When earth was subsequently added by Empedocles, man was presented with four elements from which all else was derived -- Air, Earth, Fire and Water. Comingling and separation were the causes of physical change.

The philosophical explanation for the structure of matter approached its zenith around 400 B.C. when Democritus supposed that matter consisted of hard, indivisible particles (atomos). All of these, he said, were made of the same common basic substance but differed among themselves by size and shape, and change was the result of changing their arrangements. Aristotle's explanation, however, became the final word. He varied from the Democritus conception by discounting the particulate nature of matter and by defining change as more than a rearrangement. Chemical compositions were examples of substantial change; the elements did not remain what they were, but became a compound. And every part of the compound was identical to itself. Although the controversy continued and modifications were suggested by various philosophers, Aristotle's explanation dominated science for two thousand years.

The nineteenth century ushered in the beginnings of modern atomic theory. Although it has undergone innumerable refinements and qualifications, the basic theory proposed by John Dalton in 1808 has remained amazingly constant. Matter is composed of atoms which can neither be created nor destroyed, subdivided or changed. Atoms of the same substances are alike; atoms of different substances differ from the others in weight and other characteristics. And, atoms combine in definite numerical proportions.

Enter the poet

Contemporary physics is not concerned with attempting to picture the atom; indeed it warns that we should not try to do so. But one of the fundamental conceptions of the atom is that this basic unit of the solid world is overwhelmingly empty space.

The classic piece, "Prometheus Unbound," was written by Shelley (1792-1822) during the time of Dalton's publication of his Atomic Theory.

> A sphere, which is as many thousand spheres;
> Solid as crystal, yet through all its mass
> Flow, as through empty space, music and light;
> Ten thousand orbs involving and involved,
> Purple and azure, white, green and golden,
> Sphere within sphere; and every space between
> Peopled with unimaginable shapes,
> Such as ghosts dream dwell in the lampless deep;
> Yet each inter-transpicuous; and they whirl
> Over each other with a thousand motions,
> Upon a thousand sightless axles spinning,
> And with the force of self-destroying swiftness,
> Intensely, slowly, solemnly, roll on,
> Kindling with mingled sounds, and many tones,
> Intelligible words and music wild.
> With mighty whirl the multitudinous orb
> Grinds the bright brook into an azure mist
> Of elemental subtlety, like light;

It interpenetrates my granite mass,
Through tangled roots and trodden clay doth pass
Into the utmost leaves and delicatest flowers;
Upon the winds, among the clouds 'tis spread,
It wakes a life in the forgotten dead,
They breathe a spirit up from their obscurest bowers;

Man, oh, not men! a chain of linked thought.
Of love and might to be divided not,
Compelling the elements with adamantine stress;
As the sun rules even with a tyrant's gaze
The unquiet republic of the maze
Of Planets, struggling fierce towards heaven's free wilderness.
- from "Prometheus Unbound" -
- Percy Bysshe Shelley -

    Shelley, as the communicator to the non-scientific world, described the atom vividly as solid yet void -- "solid as crystal" -- thereby creating a visual image of a hard but mystically transparent sphere not unlike glass marbles whirling and spinning. But he went further in the second and third lines by describing a phenomenon that at that time could have existed only in his own fanciful imagination. As Carl Grabo, an authority on Shelley's poetry, wrote: "Either Shelley is indulging in a flight of fancy, or, packed within these lines are recondite meanings."[1]
    Were the lines written today, "... yet through all its mass / Flow, as though empty space, music and light," quite possibly the uniqueness might go unnoticed in view of the contemporary knowledge of the atom. For within the atom, which is predominantly empty space, there is light and music -- light as the quanta emitted from the excitation of electrons, light as the gamma radiation emitted from the nucleus of the radioactively unstable atom, music as from the rhythmatic pulsations of intraatomic vibrations.
    Further, as one reads of the spheres whirling "over each other

with a thousand motions, / Upon a thousand axles spinning," one cannot reject the vision of the individually spinning electrons spinning about the nucleus in their statistically precise orbitals ... spinning within spinning. The entire mental image is that of a dynamic energy bordering on the edge of chaos, yet with a pattern of incredible orderliness. The piece brings to mind the exclamation of William Ayre written in 1752.

> What greater paradox in words can be!
> That what I see, is not the thing I see:
> - from "Truth: Counterpart to Mr. Pope's Essay"-
> - William Ayre -

In "Soul and Sense," written in 1896, poet Hannah Kimball captures the frenzied motion that Shelley described. One can accept it as a restatement of the material world -- kinetic molecular motion, celestial motion, the forces of attraction between all bodies -- yet from which the soul of man remains apart. One can be sucked into the whirlpool of influences, or by ones own "stubborn power" can escape. Man is more than a conglomerate of "motley molecules."

> Myriads of motley molecules through space
> Move round triumphant. By their whirlpool pace
> Shall we be shaken? All in earth's vast span,
> Our very bodies, veer to other shapes;
> Mid the mad dance one stubborn power escapes,
> Looks on and marvels, -'t is the soul of man.
> - from "Soul and Sense" -
> - Hannah Kimball -

We will recall that I. A. Richards has written that in some poems some words can be almost entirely missed or neglected without loss. The form "seems as an inexplicable premonition to a meaning we have not yet grasped." Consider his thought in terms of Soul and Sense." In this poem the word "molecule" is a placemarker for a multitude of provocative thoughts. Consider "Myriads of motley *ideas* ... mid the mad dance one stubborn power escapes ... The soul of man." Man is bombarded by ideas just as he is by molecules. He is shakened and shaped by them, but retains within himself the power to escape. Similarly, one could substitute for molecules other words -- religions, people, temptations -- and the message would be the same. Man is in control; he is the master of his fate; his soul can overcome.

One sometimes finds in short poems a meager collection of words which express a thesis far more profound than the literal meaning of the words themselves. Such is typified by the two-line poem "Atom from Atom."

>   Atom from atom yawns as far
>   As moon from earth, or star from star.
>   - "Atom from Atom"
>   - Ralph Waldo Emerson -

One would have to ask if Emerson (1803-1882) were simply stating a fact of relative distances. If so, for what purpose? Or rather, should we consider this poem to be asking a much more significant question? Beyond the farthest stretches of the imagination could it be that the atoms about which we talk so casually are the planets of a submicroscopic universe? And that our moon and planets are the atoms of a supermacroscopic compound and universe? Dare we to consider that perhaps the Solar System Theory of atomic structure as promulgated by Neils Bohr in 1913, with its whirling, spinning electrons could be the planets of an incredibly small cosmos? Unquestionably, such thoughts and conjectures can be discounted as unfounded romanticizing. And yet, Norman Pearson cautions that one should not discount the seeming deliberate fictions arising from the poet's mind. As one recalls former impossiblities that are now accepted as the ordinary, who can say with certainty and assurance that contemporary imaginations are not the forerunners of tomorrow's reality?

William Blake could never forgive the scientist for having analyzed the divine mystery of immediate experience into merely physically explainable elements. Though he attacked science more vigorously than any other poet of the time, still he expressed with poetic skill what the protagonists of science had been saying less effectively -- the wisdom of God is seen in nature. He would probably not have agreed that such was best revealed by the discoveries of science; but it is ironic that what some critics accept as the most refined distillation of eighteenth century scientific poetry can be observed in the first four lines of his poem "Auguries of Innocence." His entire poem is included here because we shall return to it again at which time we shall compare it with one written by Carl Sandburg.

> To see a World in a grain of sand,
> And a Heaven in a wild flower,
> Hold Infinity in the palm of your hand,
> And Eternity in an hour.
> A robin redbreast in a cage
> Puts all Heaven in a rage.
> A dove-house fill'd with doves and pigeons
> Shudders Hell thro' all its regions.
> A dog starv'd at his master's gate
> Predicts the ruin of the State.
> A horse misus'd upon the road
> Calls to Heaven for human blood.
> Each outcry of the hunted hare
> A fibre from the brain does tear.
> A skylark wounded in the wing,
> A cherubim does cease to sing.
> The game-cock clipt and arm'd for fight
> Does the rising sun affright.
> Every wolf's and lion's howl
> Raises from Hell a Human soul.
> The wild deer, wandering here and there,

Keeps the Human soul from care.
The lamb misus'd breeds public strife,
And yet forgives the butcher's knife.
The bat that flits at close of eve
Has left the brain that won't believe.
The owl that calls upon the night
Speaks the unbeliever's fright.
He who shall hurt the little wren
Shall never be belov'd by men.
He who the ox to wrath has mov'd
Shall never be by woman lov'd.
The wanton boy that kills the fly
Shall feel the spider's enmity.
He who torments the chafer's sprite
Weaves a bower in endless night.
The caterpillar on the leaf
Repeats to thee thy mother's grief.
Kill not the moth nor butterfly,
For the Last Judgment draweth nigh.
He who shall train the horse to war
Shall never pass the polar bar.
The beggar's dog and widow's cat,
Feed them, and thou wilt grow fat.
The gnat that sings his summer's song
Poison gets from Slander's tongue,
The poison of the snake and newt
Is the sweat of Envy's foot.
The poison of the honey-bee
Is the artist's jealousy.
The prince's robes and beggar's rags
Are toadstools on the miser's bags.
A truth that's told with bad intent
Beats all the lies you can invent.
It is right it should be so;
Man was made for joy and woe;
And when this we rightly know,

Thro' the world we safely go.
Joy and woe are woven fine,
A clothing for the soul divine;
Under every grief and pine
Runs a joy with silken twine.
The babe is more than swaddling-bands;
Throughout all these human lands
Tools were made, and born were hands,
Every farmer understands.
Every tear from every eye
Becomes a babe in Eternity;
This is caught by Females bright,
And return'd to its own delight.
The bleat, the bark, bellow, and roar
Are waves that beat on Heaven's shore.
The babe that weeps the rod beneath
Writes revenge in realms of death.
The beggar's rags, fluttering in air,
Does to rags the heavens tear.
The soldier, arm'd with sword and gun,
Palsied strikes the summer's sun.
The poor man's farthing is worth more
Than all the gold on Afric's shore.
One mite wrung from the labourer's hands
Shall buy and sell the miser's lands:
Or, if protected from on high,
Does that whole nation sell and buy.
He who mocks the infant's faith
Shall be mock'd in Age and Death.
He who shall teach the child to doubt
The rotting grave shall ne'er get out.
He who respects the infant's faith
Triumphs over Hell and Death.
The child's toys and the old man's reasons
Are the fruits of the two seasons.
The questioner, who sits so sly,

Shall never know how to reply.
He who replies to words of Doubt
Doth put the light of knowledge out.
The strongest poison ever known
Came from Caesar's laurel crown.
Nought can deform the human race
Like to the armour's iron brace.
When gold and gems adorn the plough
To peaceful arts shall Envy bow.
A riddle, or the cricket's cry,
Is to Doubt a fit reply.
The emmet's inch and eagle's mile
Make lame Philosophy to smile.
He who doubts from what he sees
Will ne'er believe, do what you please.
If the Sun and Moon should doubt,
They'd immediately go out.
To be in a passion you good may do,
But no good if a passion is in you.
The whore and gambler, by the state
Licensed, build that nation's fate.
The harlot's cry from street to street
Shall weave Old England's winding-sheet.
The winner's shout, the loser's curse,
Dance before dead England's hearse.
Every night and every morn
Some to misery are born.
Every morn and every night
Some are born to sweet delight.
Some are born to sweet delight.
Some are born to endless night.
We are led to believe a lie
When we see not thro' the eye,
Which was born in a night, to perish in a night,
When the Soul slept in beams of light.
God appears, and God is Light,

To those poor souls who dwell in Night;
But does a Human Form display
To those who dwell in realms of Day.
        -"Auguries of Innocence"-
        -William Blake-

    Critics have suggested that the first four lines are a denial of the impersonal mechanical view of the world held by the physicist. The poem reasserts the older view whereby the smallest pieces of the universe mirror the great universe itself. Blake appeared to subscribe strongly to the doctrine of Emanuel Swedenborg whose doctrine is capsulated so well in his statement: "The whole natural world corresponds to the spiritual world, and not merely the natural world in general, but also every particular of it."[2] But other readers interpret these lines as an affirmative of the scientific perspective.

    The remaining portion of the poem, after the first quatrain, is a long series of proverbs which give the modern day reader some insight into the culture of the time. We may wish to speculate on their relationship to the opening refrain.

    The inclusion of science appears on occasion in unusual works. Emerson uses the Sphinx to introduce the idea of kinetic molecular motion and electrostatic attractive forces. This is found enunciated clearly in the fourth stanza.

The Sphinx is drowsy,
Her wings are furled:
Her ear is heavy,
She broods on the world.
"Who'll tell me my secret,
The ages have kept? --
I awaited the seer
While they slumbered and slept: --

"The fate of the man-child,
The meaning of man;
Known fruit of the unknown;
Daedalian plan;
Out of sleeping a waking,
Out of a waking sleep;
Life death overtaking;
Deep underneath deep?

"Erect as a sunbeam,
Upspringeth the palm;
The elephant browses,
Undaunted and calm;
In beautiful motion
The thrush plies his wings;
Kind leaves of his covert,
Your silence he sings.

"The waves, unashamed,
In difference sweet,
Play glad with the breezes,
Old playfellows meet;
The journeying atoms,
Primordial wholes,
Firmly draw, firmly drive,
By their animate poles.

"Sea, earth, air, sound, silence,
Plant, quadruped, bird,
By one music enchanted,
One deity stirred,---
Each the other adorning,
Accompany still;
Night veileth the morning,
The vapor the hill.

"The babe by its mother
Lies bathed in joy;
Glide its hours uncounted,---
The sun is its toy;
Shines the peace of all being,
Without cloud, in its eyes;
And the sum of the world
In soft miniature lies.

"But man crouches and blushes,
Absconds and conceals;
He creepeth and peepeth,
He palters and steals;
Infirm, melancholy,
Jealous glancing around,
An oaf, an accomplice,
He poisons the ground.

"Out spoke the great mother,
Beholding his fear;---
At the sound of her accents
Cold shuddered the sphere:---
'Who has drugged my boy's cup?
Who has mixed my boy's bread?
Who, with sadness and madness,
Has turned my child's head?' "

I heard a poet answer
Aloud and cheerfully,
"Say on, sweet sphinx! thy dirges
Are pleasant songs to me.
Deep love lieth under
These pictures of time;
They fade in the light of
Their meaning sublime.

"The fiend that man harries
Is love of the Best;
Yawns the pit of the Dragon,
Lit by rays from the Blest.
The Lethe of Nature
Can't trace him again,
Whose soul sees the perfect,
Which his eyes seek in vain.

"To vision profounder,
Man's spirit must dive;
His aye-rolling orb
At no goal will arrive;
The heavens that now draw him
With sweetness untold,
Once found,---for new heavens
He spurneth the old.

"Pride ruined the angels,
Their shame them restores;
Lurks the joy that is sweetest
In stings of remorse.
Have I a lover
Who is noble and free?---
I would he were nobler
Than to love me.

"Eterne alternation
Now follows, now flies;
And under pain, pleasure,---
Under pleasure, pain lies.
Love works at the centre,
Heart-heaving alway;
Forth speed the strong pulses
To the borders of day.

"Dull Sphinx, Jove keep thy five wits;
Thy sight is growing blear;
Rue, myrrh and cummin for the Sphinx,
Her muddy eyes to clear!"
The old Sphinx bit her thick lip,---
Said, "Who taught thee me to name?
I am thy spirit, yoke-fellow;
Of thine eye I am eyebeam.

"Thou art the unanswered question;
Couldst see thy proper eye,
Alway it asketh, asketh;
And each answer is a lie.
So take thy quest through nature,
It through thousand natures ply;
Ask on, thou clothed eternity;
Time is the false reply."

Uprose the merry Sphinx,
And crouched no more in stone;
She melted into purple cloud,
She silvered in the moon;
She spired into a yellow flame;
She flowered in blossoms red;
She flowered into a foaming wave:
She stood Manadnoc's head.

Through a thousand voices
Spoke the universal dame;
"Who telleth one of my meanings
Is master of all I am."
                - "Sphinx" -
              - Ralph Waldo Emerson -

    His use of "primordial wholes" in line 6 of the fourth stanza is appropriate as, indeed, the atom is the first unit from which substance is derived, and the concept can be traced back to the primary material suggested by Democritus. One notes, too, his reference to the attraction which draws the atoms together yet repells them as well. The latter seems to be an apparent contradiction -- both attracting and repelling. But that contradiction may well be a clue to the poem's meaning. The Sphinx's first utterance is "Who'll tell me my secret /The ages have kept?". The last two lines, "Who telleth me one of my meanings /Is master of all I am," does not necessarily mean, as R. A. Yoder writes, "that the poet has mastered the Sphinx, but more likely that he has not been able to tell one meaning, since according to the organic view of things one cannot understand any part before he fully understands the whole. Thus the riddle is unsolved."[3] What is the ultimate source of primary material -- of the atom? What is the mechanism of creation? Why is there attractions? These, and others, are the secrets that the ages have kept. "Nature transcends man's understanding."

Questions are frequently at the heart of poetry, and the poet creates diverse ways of presenting the reader with an opportunity to extract and reflect upon them. The brief, four-line poem entitled, "Earth," is a good example.

> "A planet does not explode of itself," said drily
> The Martian astronomer, gazing off into the air-
> "That they were able to do it is proof that highly
> "Intelligent beings must have been living there."
>   - "Earth" -
>   - John Hall Wheelock -

As has been written earlier, the impact of poetry does not necessarily result from the meaning of the words per se, but from the emotions, the thoughts, the heightened awareness they evoke. What at first reading appears as an amusing rhyme, grows into a powerful realization: If we are so smart, then, why this?

Although we might prefer to look upon these lines as just humorous verse, perhaps we should consider what has been thought and written by others in a more serious vein.

> It was science that gave us the free world of the Renaisance and the apparent freedom and dignity of man. Again it is science which seems to have taken them from us as we look out now in fear and trembling.[4]
>
> -Norman Holmes Pearson-

Paul Dehn (1912-1976) uses a similar technique yet more directly satirical. Two verses of the poem, "Rhymes for a Modern Nursery," target the nuclear age and to a large extent imply a science that is oblivious to its impact upon mankind. Is the threat to man the result of the products of the scientific enterprise, or the result of the disinterest and lack of concern of the scientist? Whereas the aestheticism, "art for art's sake" is well known and familiar, does there exist with the scientist, "science for science's sake"?

> Hey diddle diddle,
> The physicists fiddle,
> The bleep jumped over the moon.
> The little dog laughed to see such fun
> And died the following June.
>
> Jack and Jill went up the hill
> To fetch some heavy water.
> They mixed it with the dairy milk
> And killed my youngest daughter.
>
> Two blind mice
> See how they run!
> They each ran out of the lab with an oath,
> For the scientist's wife had injected them both
> Did you ever see such a neat little growth
> On two blind mice?

Little Miss Muffet
Crouched on a tuffet,
Collecting her shell-shocked wits.
There dropped (from a glider)
An H-bomb beside her----
Which frightened Miss Muffet to bits.
>   -"Rhymes for a Modern Nursery"-
>   -Paul Dehn-

Wherner Von Braun (1912-1976), the German rocket developer and later reputed to be the father of the American space program, was once asked why he worked for Nazi Germany to develop the V-2 rocket as an instrument of death and destruction. His reply was that it was the only way to advance the technology of rocketry and ultimately space travel.

Science and poetry are sometimes distinguished by the way each approaches phenomena. The scientist defines the law and explains it; he defines it exactly and unequivocally, so that the phenomenon is unquestionably unambiguous. The poet and his reader, once the phenomenon has been expressed in a significant form, appreciate it for its own sake as a unique experience. As such, some critics believe that science has performed an inestimable service to poets in forcing them by a redefinition of physical reality to search out a revitalized manner of expression within the boundaries of truth.

Consider Newton's universal gravitational attraction referred to as the Third Law of Motion:

Any two material objects anywhere in the universe are attracted toward one another with a force (F) which is proportional to the mass (m) of each object and inversely proportional to the square of the distance (s) between them.

This gravitational force can be expressed symbolically as:

$$F = \frac{m_1 m_2}{s^2}$$

With this relationship firmly in mind, our attention is called to Francis Thompson's (1859-1907) poem, "Mistress of Visions".

> When to the new eyes of thee
> All things by immortal power
> Near or far,
> Hiddenly
> To each other linked are
> That thou canst stir a flower
> Without troubling a star.
> -from "Mistress of Visions"-
> -Francis Thompson-

We may consider this work as no more than a statement of mechanical fact just as when Newton theorized that as the apple fell toward the earth, the earth also moved toward the apple. This poem says that according to Newton's third law, if one uproots a flower, then a star is displaced accordingly.

On the other hand, we may find an even deeper and more reassuring meaning. The universe, as created by God, is inextricably interwoven. No body is too small, no body is too inconspicuous, for its presence to be unfelt or to go unnoticed by the Almighty. Thompson used this poem as another way of expressing his belief in God's concern for man.

Twentieth century poet John Updike incorporated a visual image of an elementary science demonstration into his poem "The Great Scarf of Birds." The first verse sets the tone of late autumn by its description of the trees, some of which are full of color and some already barren, and the migration of birds of all types.

As if out of the Bible or science fiction,
a cloud appeared, a cloud of dots
Ripe apples were caught like red fish in the nets
of their branches. The maples
were colored like apples,
part orange and red, part green.
The elms, already transparent trees,
seemed swaying vases full of sky. The sky
was dramatic with great straggling V's
of geese streaming south, mare's-tails above them;
their trumpeting made us look up from golf.
The course sloped into salt marshes,
and this seemed to cause the abundance of birds.

like iron filings which a magnet
underneath the paper undulates.
It dartingly darkened in spots,
paled, pulsed, compressed, distended, yet
held an identity firm: a flock
of starlings, as much one thing as a rock.
One will moved above the trees
the liquid and hesitant drift.

Come nearer, it became less marvelous,
more legible, and merely huge.
Like dutiful Lot, I turned my back.
Later, I lazily looked around.

The rise of the fairway above and behind us was tinted,
so evenly tinted I might not have noticed
but that at the rim of the delicate shadow
the starlings were thicker and outlined the flock
as an inkstain in drying pronounces its edges.
The gradual rise of green was vastly covered;
I had thought nothing in nature could be so broad but grass.

And as
I watched, one bird,
prompted by accident or will to lead,
ceased resting; and, lifting in a casual billow,
the flock ascended as a lady's scarf,
transparent, of gray, might be twitched
by one corner, drawn upward and then,
decided against, negligently tossed toward a chair.
Melting all thought, the southward cloud withdrew into the air.

- "The Great Scarf of Birds" -
- John Updike -

 The second verse recalls to the memory of the reader the time he first saw the black filings of iron spread upon a piece of paper. He recalls the symmetrical patterns created as if by magic by the unseen magnetic forces beneath. The iron metal mass would move and undulate, not by choice but by the invisible force which directed it and over which it had no control. Within the force, the individual ceased to exist as such. The mass of iron, the flock of birds, flowed, spread, contracted, settled, then rose again as a single fluid will. Thus Updike has used a phenomenon of physical science to imply within the poem a suggestion of forces, whether they be magnetic, instinctive, or spiritual, that establish patterns; unseen forces act upon all.

 One should not depart the consideration of "attractions" without examining Emmett Williams' mid twentieth century poem "Like Attracts Like". In this poem he creates a purely visual message from the physical structure of the words. Poet Ernest Jandl writes: "There must be an infinite number of methods of writing experimental poems, but I think the most successful are those which can be used only once, for the result is a poem identical with the method by which it is made. The method used again would turn out exactly the same poem." Williams' own commentary on his poem is that the "poem says what it does, and does what it says."[5] Could Newton have expressed his law better?

```
like    attracts   like
like    attracts   like
like    attracts   like
like    attracts   like
like    attracts   like
like    attracts   like
like   attracts  like
likeattractslike
likeattractlike
likettraclike
likttraclike
liktralike
liktlikts
```

-"Like Attracts Like"-
-Emmett Williams-

CHAPTER III ORDER

> Nobody, I think, ought to read poetry ... who cannot find a great deal more in them than the poet ... has actually expressed.
> -from *The Marble Faun*-
> -Nathaniel Hawthorne-

The perspective of order

The belief since ancient times has been that in the beginning all was chaos and that the Creator brought forth order, and continues to bring forth more order. For the natural philosopher, the evidence was indisputable that God himself was the principle of order who separated light from dark, earth from sea, and who patterned the heavenly pathways. Thus to bring order from chaos was divine and was, therefore, a moral imperative for all.
    Aristotle's idea of order was central to his philosophy and was based upon his acceptance of the natural celestial order. Plato, on the other hand, had searched for an overt demonstration of intelligibility in things. He had searched for patterns, rules, and constancy to substantiate his correlation of deity, man, and order.

He sought immutable and unchangeable forms as the evidence of a maturity of orderliness. We note a similarity between the idea of Plato's constancy and the requisite of the primary material -- the atom.

Even as the Greeks contemplated chaos and order, similarly on the other side of the globe the early Chinese philosophers suggested the same. The cosmogony of the Huai nan tzu began with nothing of physical shape --- homogeneous nothingness -- which writers today would define as statistical randomness or complete chaos.

> There is a thing confusedly formed,
> Born before heaven and earth.
> Silent and void
> It stands alone and does not change,
> Goes round and does not weary,
> It is capable of being the mother of the world.
> I know not its name
> So I style it "the way."
> ....
> Man models himself on earth,
> Earth on heaven,
> Heaven on the way,
> And the way on that which is naturally so.[1]
>   - from *Lao-tzu,* ch. 25 -
>   -Lao Tzu -

The cosmic order of Aristotle and the teleological order of Plato dominated western philosophical thought until the seventeenth century mechanical order, introduced by Newton, made its appearance. Newton in no way deliberately degraded the conception that God created an orderly universe. From the point of view of himself and others, his scientific/mathematical proof of universal order gave a new meaning to the earlier view of God's creation. Francis Webb's, "Poems," summarized what the new science enabled the philosopher to see: "Tracing through all his works th' Almighty hand." Typical of this recognition of God's majestic universe and his wonderful creation was the poem by Jane Brereton in 1735. In "Poems on Several Occasions," she praised Newton for revealing God through order.

> Newton, th' All-wise Creator's work explores,
> Sublimely, on wings of knowledge, soars;
> The establish'd order, of each orb, unfolds,
> And th' omnipotent God, in all, beholds:
> If to the dark abyss, or bright abode,
> He points; the view still terminates in God.
>   - from "Poems on Several Occasions" -
>   - Jane Brereton -

Lest we be led to believe that Newtonian mechanical order was accepted by all as a reaffirmation of God's order, then we need only to reflect upon a portion of Pope's "Essay on Man":

> Go, wonderous creature! Mount where science guides,
> Go, measure earth, weigh air, and state the tides;
> Instruct the planets in what orbs to run,
> Correct old Time, and regulate the Sun;
> Go, teach Eternal Wisdom how to rule -
> Then drop into thyself, and be a fool!
>   - from "Essay on Man" Epistle II -
>   - Alexander Pope -

During all of the eighteenth century, poets, philosophers, clergy, and scientists were attempting to reconcile within and among themselves the subtleties of order. If order were God's work, and God was good and merciful, how could one account for the natural disasters that fell upon man and the unpredictability of nature, both of which were obviously disruptions to an orderly system? The poem, "Truth: Counterpart to Mr. Pope's Essay on Man," addressed the dilema.

> What greater paradox in words can be!
> That what I see, is not the thing I see:
> That groans, and shrieks, and screams, and dying cries,
> Are mixed with something else, great harmony;
> That all monstrous crimes, which men commit,
> Are universal order, just and fit:
> That earthquakes, whirlwinds, deluges, and flames,
> All, all are order under other names.
>   - from "Truth: Counterpart to
>   Mr. Pope's Essay on Man" -
>   - William Ayre -

"Truth ..." is an expression by Ayre of his supreme faith that the natural disasters were orderly components of the orderly universe. He had no data to support his belief nor to explain the origin and cause of whirlwinds, hurricanes, blizzards, and droughts, much less the genesis of earthquakes. Superstition and mysticism had been adequate explanations for thousands of years. Knowledge of the jet stream was far removed, and the idea of vulcanism and plate tectonics was not to come for another three centuries. Ayre was calling upon the reader to accept the paradox that apparent disorder is "... order under other names."

In a similar vein, Emerson also called upon the reader to accept apparent disorder. He wrote that they "Deceive us, seeming to be many things, / And are but one." He was convinced that through logical reasoning the oneness was obvious.

> By fate, not opinion, frugal Nature gave
> One scent to hyson and to wall-flower,
> One sound to pine-groves and to waterfalls,
> One aspect to the desert and the lake.
> It was her stern necessity: all things
> Are of one pattern made; bird, beast, and flower,
> Song, picture, form, space, thought, and character
> Deceive us, seeming to be many things,
> And are but one. Beheld far off, they part

> As God and devil; bring them to the mind,
> They dull its edge with their monotony.
> To know one element, explore another,
> And in the second reappears the first.
> The specious panorama of a year
> But multiplies the image of a day,---
> A belt of mirrors round a taper's flame;
> And universal Nature, through her vast
> And crowded whole, an infinite paroquet,
> Repeats one note.
>
>   - "Xenophanes" -
>   - Ralph Waldo Emerson -

Still the issue of seeming disorder and disaster within God's order posed the philosophical question of good and evil. If order were good because order was God, then what was the place of evil? Pope used his "Essay on Man," to pronounce that all evils, including the harshness of nature, were not only consistent but also were a necessary part of God's infinite wisdom. The last six lines summarize his argument.

> All nature is but Art unknown to thee;
> All Chance, Direction, which thou canst not see;
> All Discord, Harmony not understood;
> All partial Evil, universal Good:
> And spite of pride, in erring Reason's spite
> One truth is clear, *Whatever is, is right.*
>
>   - from "Essay on Man" Epistle I -
>   - Alexander Pope -

To a large measure, seventeenth century poets wrote lavishly about the order of the universe. In some cases the praise was recognition that the order proven by the mechanistic philosophy was the reinforcement of one's belief that earth, and indeed the universe, was God's creation. John Dryden in the late 1600's wrote such an acclaim.

> This as a piece too fair
> To be the child of Chaos, and not of Care.
> No atoms casually together hurl'd
> Could e'er produce so beautiful a world.
>   - from "Design" -
>   - John Dryden -

A similar poem was written by John Addison (1672-1719) slightly later. The handiwork of God and the importance of man were reaffirmed as he wrote that the evenings' celestial show "Confirm the tidings ... / and spread the truth ...", and "... utter forth a glorious voice; / 'The Hand that made us is divine.'"

> The spacious firmament on high,
> With all the blue etheral sky,
> And spangled heavens, a shining frame,
> Their great Original proclaim.
> The unwearied Sun, from day to day,
> Does his Creator's power display;
> And publishes to every land
> A work of an Almighty hand.
>
> Soon as the evening shades prevail,
> The Moon takes up the wondrous tale;
> And nightly to the listening Earth
> Repeats the story of her birth:
> Whilst all the stars that round her burn,
> And all the planets in their turn,
> Confirm the tidings as they roll
> And spread the truth from pole to pole.
>
> What though, in solemn silence, all
> Move round the dark terrestrial ball?
> What though nor real voice nor sound
> Amidst their radiant orbs be found?

> In Reason's ear they all rejoice,
> And utter forth a glorious voice;
> Forever singing as they shine,
> "The Hand that made us is divine."
>   - "Hymn" -
>   - Joseph Addison -

The concept of order found another use apart from glorifying the Creator. Because of the divine origin and embodiment in the cosmological conception of the Chain of Being, order was used as a weapon against social discontent and especially against all equalitarian movements. Human society was considered to be well constituted as long as it corresponded to the grand order. Referring once again to Pope's "Essay on Man," we find him expressing this moral rightness of social strata. An individual's place in society was to be recognized as a part of God's plan, and, in essence, it was the individual's duty to accept his place without challenge.

> Order is Heav'ns first law; and this confest,
> Some are, and must be, greater than the rest,
> More rich, more wise.
>   - from "Essay on Man" -
>   - Alexander Pope -

Thus agitation for social change was contrary to God's will. "The universe resembles a large and well-regulated family, in which all of the officers and servants, and even the domestic animals, are subservient to each other in the proper subordination;" wrote Soame Jenyns, philosopher and essayist of the 1700's. "Each enjoys the priviledges and perquisites peculiar to his place, and at the same time contributes, by that just subordination, to the magnificence and happiness of the whole." But before we become too appalled at the callous rationalization for a continuation of social stratefication or castes, we should examine the teachings of seventeenth century Christianity. The

New Testament was repleat with instructions that one should function within the system.

> 22. Wives, submit yourselves unto your own husbands, as unto the Lord.
> 23. For the husband is the head of the wife, even as Christ is the head of the church; and He is the saviour of the body.
> 24. Therefore as the church is subject unto Christ, so let the wives be to their own husbands in everything.
> - Ephesians 5 -

> 5. Servants be obedient to them that are your masters according to the flesh, with fear and trembling, in singleness of your heart, as unto Christ.
> 9. And, ye masters, do the same thing unto them, forbearing threatening; knowing that your Master also is in heaven; neither is their respect of persons with him.
> - Ephesians 6 -

> 5. Blessed are the meek, for they shall enherit the earth.
> 17. Think not that I am come to destroy the law or the prophets; I am not come to destroy, but to fulfill.
> - St. Matthew 5 -

> 19. Lay not up for yourselves treasures upon earth, where moth and rust doth corrupt, and where thieves break through and steal;
> 20. But lay up for yourselves treasures in heaven, where neither moth nor dust corrupt, and where thieves do not break through nor steal.
> - St. Matthew 6-

> 25. And he said unto them, Render therefore unto

Caesar the things which be Caesar's, and unto God
the things which be God's.
                - St. Luke 20 -

Disorder

For some time now our attention has been directed toward order and the orderly universe. From the original chaos the universe was made orderly, and a belief is that it continues to become more orderly. It is now time to examine another concept, the concept of disorder -- which is named entropy.

Like all concepts of science, entropy evolved. It was not the instant revelation of one particular person at some isolated point in time. In the early 1800s, a French army officer, Sadi Carnot, was studying the steam engine to determine how its efficiency could be improved. His conclusion was that greater energy was available for use if two temperature extremes could be maintained -- high heat input and a lower heat output. The greater the difference between the hot component of the machine and the cold component, the more energy there would be available. Today, we look upon this generalization as a very logical conclusion. When something hot becomes cold, the energy that is released is available for use by us, or in reality the machine. We call this free energy or available energy.

The concept of the high energy level and the low energy level can be visualized as a mechanical phenomenon also. As an object falls, it possesses kinetic energy. The farther it falls before stopping, the more energy it possesses. In spite of the logic, however, it was thirty years before Rudolf Clausius (1822-1888) verbalized these observations as the Laws of Thermodynamics.

Clausius' first law stated that all matter and energy in the universe is constant; it cannot be created nor destroyed, but it can be changed. Hydrogen and oxygen can become hydrogen oxide (water), liquid water can become solid; and ice can melt. The second law, which is called the entropy law, stated that matter and energy can change spontaneously, but in only one direction -- from high energy to low energy. A cup of hot coffee (high heat energy) becomes cold (low heat energy); a cup of cold coffee never begins to boil spontaneously.

Clausius went further to describe the energy phenomena as order. Creating order was another way of storing energy. Building a skyscraper is a process of creating order. The random grains of sand, pieces of metal, strands of wire, piles of gravel, and pools of water are assembled into an orderly arrangement and as such represent increased energy. Just the act of placing a brick at the sixty-fifth level increases the potential energy of the brick, and the building. If the brick were to fall from its perch at the sixty-fifth floor and land on your foot, then you would be aware of its energy content because then it would have been transferred to the foot -- smashing and crushing the bones and flesh. And that was Clausius' observation. We expect the brick to fall and destroy (disorder) the foot. We do not expect a smashed foot suddenly to reconstruct itself, and for a brick spontaneously to hurl itself upwards sixty-five floors to a precisely predetermined spot. Clausius' generalization, then, was that order is abnormal and that disorder is the ultimate conclusion -- therefore, normal; the universe began with structure and order. His second law of thermodynamics says that everything in the universe is irrevocably moving in the direction of disorder and random chaos. Entropy is increasing.

Theodore Spencer (1902- 1949) viewed entropy as a depressing concept. It told him that the direction of existence is not towards a more perfect, orderly universe, but toward chaos and disorder. He attempted, however, to extract an element of hope from the inevitable conclusion.

> Matter whose movement moves us all
> Moves to its random funeral,
> And Gresham's Law that fits the purse
> Seems to fit the universe.

>Against the drift what form can move?
>(The God of order is called Love.)
>- "Entropy" -
>- Theodore Spencer -

It is interesting to note how he describes his disenchantment further by drawing the parallel of Gresham's Law -- bad money drives out good -- with entropy -- disorder is inevitable. On the other hand, his message is clear; love overcomes all.

A singular observation is that the modern world, although aware of the concept of entropy for a hundred years, is still willing to see Creation as changing chaos into a progressively higher state of order. But if we go back in time to the period between Newton's order and Clausius' disorder, we will find some poets who did in fact anticipate an ultimate creation in reverse. Aaron Hill, in 1721, was one such poet.

>The stars forget their laws, and like loose planets stray.
>See how their elements resign
>Their numerous charge, their scatter'd atoms home repair,
>Some from the earth, some from the sea, some from the air.
>- from "The Judgment Day" -
>- Aaron Hill -

Another poem representing the same general topic was written by Elizabeth Rowe (1674-1737). It describes the complete disruption of the Newtonian mechanical model. We would have difficulty today deciding whether her poem was a hostile reaction to the mechanical model, or whether through some intuition she viewed the order of the universe as being only momentary. At the instant of God's creation the universe was the perfection of order, but from that instant forward it had begun to deteriorate, to unwind, to move toward the inevitable chaos. A portion of her poem describes the pathway dramatically.

> "And now begins the universal wreck;
> The wheels of nature stand, or change their course,
> And backward hurrying with disorder'd force,
> The long establish'd laws of motion break ...
> Now mightier pangs the whole creation feels;
> Each planet from its shelter'd axis reels,
> And orbs immense on orbs immense drop down,
> Like scatt'ring leaves from off their branches blown."
> - from "The Conflagration: An Ode" -
> - Elizabeth Rowe -

If, in fact, her intuition was that disorder was the fate of the universe, she had just cause. We must remember that in her time few persons questioned the perfection of God and, therefore, the perfection of his creation. This belief, however, was tempered by the doctrine of original sin. Everything was perfect until man came on to the scene. From that point onward, his original sin precluded the possibility of man ever reaching a state of perfection, and, therefore, it could be reasoned that the same would hold true for the universe itself. Did not the falling of stars indicate that the order was becoming unraveled?

One other aspect of the entropy law should be mentioned. If total energy can be changed spontaneously in only one direction, from high to low, from hot to cold, the time will come when all energy reaches the same lower level. An equilibrium will be established to the extent that no energy differences exist anywhere. In the mid 1800s, Herman Helmholtz, a German physicist, coined the phrase, "heat death," to describe the condition whereby the universe is gradually running down and will eventually reach the point of maximum entropy (disorder). At that point, all available energy will have been consumed and no more activity can occur.

With "heat death" in mind, let us look at the poem "Fire and Ice."

> Some say the world will end in fire,
> Some say in ice.
> From what I've tasted of desire
> I hold with those who favor fire.
> But if I had to perish twice
> I think I know enough of hate
> To say that for destruction ice
> Is also great
> And would suffice.
>
>   - "Fire and Ice" -
>   - Robert Frost -

    John Doyle, a biographer and analyst of Frost and his poetry, recognizes that the literal reader will want to know the answer to which will end the world, Fire or Ice? In the tradition of Frost's other "unanswered question" poems, we are left to decide for ourselves. However, the poem is not without its conclusions. Doyle writes: "The poem offers two insights: one, that what brings the end is unimportant -- the important fact is the end itself; second, it identifies two of the destroyers of life (desire and hate), destroyers that are not usually associated with the end, or ever recognized as being destroyers."

    If we, who now know something of entropy and the theory of "heat death," are asked for our evaluation of Frost's choices for the end of the world, we can recognize the possibility of his scenario of ice occurring. But in the twenty years following his writing of the poem, man was launched into a new age by the atomic bomb. As Issac Asimov has said, "Fire may get us yet."

    A second poem by Frost should be examined within the context of entropy.

> Nature's first green is gold,
> Her hardest hue to hold.
> Her early leaf's a flower;
> But only so an hour.

> Then leaf subsides to leaf.
> So Eden sank to grief,
> So dawn goes down to day.
> Nothing gold can stay.
> - "Nothing Gold Can Stay" -
> - Robert Frost -

Professor Charles Anderson describes the poem as "running downhill from the beginning." The leaf diminishes, the dawn goes down, Eden sinks, and everything seems to be declining from the pinnacle of nature and human history. The Creation was more glorious than the finished product, and gold - the color-word that dominates the poem -- is the emblem of that glory. He believes, however, that Frost is insisting that gold "cannot stay or be stayed." In Anderson's mind, the poem reinforces the scientific thought that "points to a universe running down rather than up." Without the word, entropy, being used, he believes that the implication of inevitable deterioration is a just conclusion.

We must look again at the precise words of the Laws of Thermodynamics before we accept or reject what has been said about entropy. Whereas the first law refers to the universe -- that all matter and energy is constant -- the other laws describe a "closed system." (Remember that Carnot's work which led to the laws was based upon the steam engine.) The entropy law seems to be correct when applied to closed systems. But the question which separates the scientists and the philosophers, even from among themselves, is the extent to which the universe is closed or is limitless. If it is enclosed, then it is a very large enclosure. If it is limitless, then it is very difficult for man to comprehend something without any boundaries at all. As Jeremy Rifkin wrote in 1980 in *Entropy*:

> ... some will remain unconvinced that there are physical limits that place restraints on human action in the world. Others will be convinced but will conclude with

despair that the Entropy Law is a giant cosmic prison from which there is no escape. Finally, there will be those who see the Entropy Law as the truth that can set us free. The first group will continue to uphold the existing world paradigm. The second group will be without a world view. The third group will be harbingers of the new age.[5]

- from *Entropy* -
- Jeremy Rifkin -

Now that we have some knowledge of order and disorder, of entropy and energy, let us examine the ways a poet explains the concepts. You will recall that poets have long been interpreters between science and the ordinary man. Sometimes the poets simplify; sometimes they cloud; sometimes they opine; sometimes they leap beyond. As you read the "First and Second Law," try to decide what is being done by the poet.

And all because of the Second Law of Thermodynamics
That lays down ...
That you can't pass heat from a cooler to a hotter.
You can try if you like, but you'd far better notta,
'Cause the cold in the cooler
Will get hotter as a rule-a,
'Cause the hotter body's heat will-a pass to the cooler,
And that's a physical law.

Oh, I'm hot.

Hot? That's because you've been Working.

That's the First and Second Law of Thermodynamics.

The First Law of Thermodynamics:
Heat is Work and Work is Heat.

The Second Law of Thermodynamics:
Heat cannot of itself pass from one body to a hotter body.

Heat won't pass from a cooler to a hotter.
You can try if you like, but you far better notta.
Thus the cold in the cooler
Will get hotter as a rule-a,
Because the hotter body's heat will pass to the cooler.

Heat is Work and Work is Heat
And Work is Heat and Heat is Work.

Heat will pass by Conduction;
Heat will pass by Convection;
Heat will pass by Ra-diation;
And that's a physical law.

Heat is Work and Work's a curse
And all the Heat in the universe
'S gonna c-o-o-l down;
'Cause it can't increase.
Then there'll be no more work
And there'll be perfect peace.

Really?

Yeah, that's entropy, man.
- "First and Second Law" -
- Michael Flanders (1922-1975) -

In this chapter we began by looking at the universe and discovering an amazing orderliness. This orderliness has been defined and verified through science, religion and poetry. We have looked at disorder, which also has been defined and verified by science. The common thread throughout the fabric of our exploration has been energy. Energy makes the world go round, literally. The poet has treated energy as he has traditionally treated any subject. Sometimes he merely describes it as though he were a scientist, albeit in a quite different language. Other times he will use the concept of science to make a point tangential to the obvious.

An example of the descriptive approach is the poem, "The Action of Electricity," by Erasmus Darwin (1731-1802) in the late eighteenth century. As we read it, we should attempt to visualize the images that Darwin constructs and to place ourselves in those times. The poem falls somewhere between the mechanical and the mystical -- mystical in its avoidance of a technical explanation of why or how, but mechanical in its implications of processes and equipment.

> Nymphs! your fine hands ethereal floods amass
> From the warm cushion, and the whirling glass;
> Beard the bright cylinder with golden wire,
> And circumfuse the gravitating fire.
> Could from each point cerulean lustres gleam,
> Or shoot in air the scintillating stream.
> So, borne on brazen talons, watch'd of old
> The sleepless dragon o'er his fruits of gold;
> Bright beam'd his scales, his eye-balls blazed with ire,
> And his wide nostrils breath'd inchanted fire.
> 'You bid gold-leaves, in crystal lantherns held,
> Approach attracted, and recede repell'd;
> While paper-nymphs instinct with motion rise,
> And dancing fauns the admiring Sage surprize.
> Or, if on wax some fearless Beauty stand,
> And touch the sparkling rod with graceful hand;
> Through her fine limbs the mimic lightnings dart,
> And flames innocuous eddy round her heart;
> O'er her fair brow the kindling lustres glare,
> Blue rays diverging from her bristling hair;
> While some fond Youth the kiss ethereal sips,
> And soft fires issue from their meeting lips.

> So round the virgin Saint in silver streams
> The holy Halo shoots it's arrowy beams.'
> - "The Action of Electricity" -
> - Erasmus Darwin -

A digression would be appropriate with this mention of electricity. On the other side of the world, science in China and other regions of the orient lagged behind that of the western world. Although significant technological innovations were developed in the earlier centuries of the Chinese civilization -- paper, abacus, metallurgy, and gunpowder -- they were more representative of ingenious inventions than of steps in the sequential development of scientific thought.

Many writers have theorized about the reasons that science did not develop and flourish in China as it did in the west. They generally agree that a dominant reason was the overriding philosophical premise of the Chinese of natural harmony. No clear demarcation existed between natural laws and human laws, between events caused by Nature and those caused by Man, and between Nature and human society. Consequently, basic ideas and explanations were never seriously challenged.

> Among creatures some lead and some follow.
> Some blow hot and some blow cold.
> Some are strong and some are weak.
> Some may break and some may fall.
> Therefore the Sage discards the extremes, the
>   extravagant, and the excessive.
> - from *Tao-te ching*, ch.29 -
> - Lao Tzu -

This portion of chapter 29 of *Tao-te ching*, or *Lao-tzu* is commonly interpreted to mean that the nature of existence is spontaneity. Events and phenomena occur because it is their way. (Lao-tzu means "the way.") They should be followed, but not interfered with. "The Sage understands Nature perfectly and knows clearly the conditions of all things. Therefore he goes along with them. He removes their delusions and eliminates their doubts. Hence people's minds are not confused and things are contented with their own nature."6

With this as the background, it is easier for us to understand how, even in the late 1800s, Chinese ideas about electricity differed dramatically from those of the west.

> Electricity is not confined to space, for there is nothing which it does not integrate and penetrate. The brain is one of the substances in which electricity assumes physical form and solid substance. Since the brain is electricity with physical form and solid substance, then electricity must be brain without physical form or solid substance. Since men know that it is the power of the brain that pervades throughout the five sense organs and the hundred bones that make them one body, they should know that the power of electricity pervades throughout heaven, earth, ten thousand things, the self and the other, and makes them one body ... Electricity is everywhere. It follows that the self is everywhere.
> - from "Ether and Humanity" -
> - T'an Ssu-t'ung -

In the poem mentioned earlier, "First and Second Law," Flanders refers to the transferring of heat (energy) by conduction, convection, and radiation. These would seem to be peculiar inclusions in a poem, but, then, nothing excapes the poet's attention. Within the entire energy phenomena a major consideration deals with wave motion. Whereas we think of wave motion in terms of water action, that is only one small example. In general, wave motion refers to a method of transferring energy from one body to another as though through a fluid environment.

If we shake up and down the end of a horizontal rope, waves are created that seem to move from the shaken end to the other. The illusion is that the rope is moving laterally, but it is not. Only the energy moves laterally; the rope does not.

In wave motion, a body absorbs vertical energy and transmits it horizontally. If we disregard any winds that may be blowing and watch a cork bobbing on the wave, it becomes apparent that the water is actually moving only up and down, not horizontally, because the cork, except for going up and down, remains basically in the same position. Ralph Waldo Emerson used this manifestation in his classic essay, "Self Reliance."

> Society is a wave. The wave moves onward, but the water of which it is composed does not. The same particle does not rise from the valley (diagonally) to the ridge. Its unity is only phenomenal.
> - from "Self Reliance" -
> - Ralph Waldo Emerson -

The movement of the wave is an illusion. The energy is transferred without the extreme lateral dislocation of the individual particles. Emerson likened this to society. Energy (social change) is transferred through the society providing great movement, action and force, yet the people as individuals do not move or change to the same extent. In other words, society goes farther and moves more than the individual people who comprise it.

The contemporary poet, May Swenson, tried to capture the action of the wave in her poem "How Everything Happens (Based on a study of the Wave)." Notice how the words, the meaning, and the visual format all contribute to the message.

Albert Einstein, the little known patent clerk, burst upon the twentieth century scene with his famous theories of relativity. In these he developed the most simplistic relationship among energy, matter, and time that is yet to be expressed. We refer of course to, $E=mc^2$: in which E is energy; m is the mass of the matter involved; and c is the velocity of light.

Although we shall consider relativity in more detail in a later

```
                                                happen.
                                            to
                                        up
                                  stacking
                              is
                      something
When nothing is happening

When it happens
              something
                      pulls
                          back
                              not
                                to
                                  happen.

When                              has happened.
      pulling back      stacking up
                  happens


        has happened                              stacks up.
When it          something                  nothing
                            pulls back while

Then nothing is happening.


                                      happens.
                                  and
                          forward
                    pushes
                  up
              stacks
      something
Then
```

         -"HOW EVERYTHING HAPPENS
         (Based on a study of the Wave)"-
         -May Swenson-

chapter, the equation momentarily calls it to our attention. An incorrect and superficial interpretation of relativity is that nothing is constant; nothing is true; there is no certainty. (This latter probably originates as a combination of relativity and the Heisenberg Uncertainty Principle.) Although these interpretations are not correct, just the word, "relativity," implies vagueness. Indeed, for many persons the thoughts expressed by the poet Morris Bishop (1895-1973) echo their feelings of uncertainty brought about by relativity and its vagueness. Yet, by clutching to the simplistic representation of the ultimate relationship among energy, matter, and time, one can infer that truth and order do exist. For Einstein also said that "God did not play dice with the universe."

> What was our trust, we trust not,
> What was our faith, we doubt;
> Whether we must or not
> We may debate about.
> The soul, perhaps, is a gust of gas
> And wrong is a form of right-
> But we know that Energy equals Mass
> By the Square of the Speed of Light.
>
> What we have known, we know not,
> What we have proved, abjure.
> Life is a tangled bowknot,
> But one thing still is sure.
> Come, little lad; come, little lass,
> Your docile creed recite:
> "We know that Energy equals Mass
> By the Square of the Speed of Light."
> - "$E=MC^2$" -
> - Morris Bishop -

Apart from the pure descriptions of scientific phenomena, poets are not disinclined to using science to bring special meanings to their works. T. S. Eliot (1888-1965) has done that with his poem "The Hollow Men." As you read it you will be aware of the despair that some critics have read into the lines. Everett Gillis' opinion is that the poem "mitigates against any possibility of spiritual progress by the hollow men." He considers this to be a modern limbo in which the soul exists totally void of human meaning.[8] Yet other writers, such as Lawrence Ryan and Friedrich Strothmann, although agreeing that in hollowness lies despair, see that emptiness is a condition of hope; they distinguish a difference between hollow and empty. They believe that Eliot associates "something positive and greatly to be desired with the conditions of emptiness."[9] With these views in mind, we need to look in particular at the stanza beginning at line 11.

>*MISTAH KURTZ---he dead.*
>A penny for the Old Guy.

>I
>We are the hollow men
>We are the stuffed men
>Leaning together
>Headpiece filled with straw. Alas!
>Our dried voices, when
>We whisper together
>Are quiet and meaningless
>As wind in dry grass
>Or rats' feet over broken glass
>In our dry cellar

>Shape without form, shade without colour,
>Paralysed force, gesture without motion;
>Those who have crossed
>With direct eyes, to death's other Kingdom
>Remember us-if at all-not as lost
>Violent souls, but only
>As the hollow men
>The stuffed men.

II

Eyes I dare not meet in dreams
In death's dream kingdom
These do not appear:
There, the eyes are
Sunlight on a broken column
There, is a tree swinging
And voices are
In the wind's singing
More distant and more solemn
Than a fading star.

Let me be no nearer
In death's dream kingdom
Let me also wear
Such deliberate disguises
Rat's coat, crowskin, crossed staves
In a field
Behaving as the wind behaves
No nearer--

Not that final meeting
In the twilight kingdom

III

This is the dead land
This is cactus land
Here the stone images
Are raised, here they receive
The supplication of a dead man's hand
Under the twinkle of a fading star.

Is it like this
In death's other kingdom
Waking alone

At the hour when we are
Trembling with tenderness
Lips that would kiss
Form prayers to broken stone.

IV

The eyes are not here
There are no eyes here
In this valley of dying stars
In this hollow valley
This broken jaw of our lost kingdoms

In this last of meeting places
We grope together
And avoid speech
Gathered on this beach of the tumid river

Sightless, unless
The eyes reappear
As the perpetual star
Multifoliate rose
Of death's twilight kingdom
The hope only
Of empty men.

V

Here we go round the prickly pear
Prickly pear prickly pear
Here we go round the prickly pear
At five o'clock in the morning.

Between the idea
And the reality
Between the motion
And the act
Falls the Shadow

       For Thine is the Kingdom

Between the conception
And the creation
Between the emotion
And the response
Falls the Shadow
       Life is very long

Between the desire
And the spasm
Between the potency
And the existence
Between the essence
And the descent
Falls the Shadow
       For Thine is the Kingdom

For Thine is
Life is
For Thine is the

This is the way the world ends
This is the way the world ends
This is the way the world ends
Not with a bang but with a whimper.
      - "The Hollow Men" -
      - T. S. Eliot -

  You will remember that the poet selects words that have a spectrum of meanings and it is the reader who extracts his own meaning from this spectrum. Thus, just as a painting can mean different things to different people, so can a poem. If we focus upon "paralyzed force," we may think of it as potential energy. Potential energy is a paralyzed force -- inactive yet poised. Potential energy is "gesture without motion" -- poised yet inactive.

From this we could interpret the meaning to be that, even with hollow men, there exists potential. It is their unknowing of this potential, and their failure to act that causes despair.

Section V begins with the entreaty to the hollow men ... "Between the idea / And the reality / Between the motion / And the act / Falls the Shadow" ... stands the hollow men. They are the potential energy; they are the gesture without motion. They are the only separation between idea and reality, between motion and action. In this regard the hollow men is everyman. As such, other men have been positioned similarly. What was their paralyzed force -- an uplifted sword ... a mighty army ...? Could it have been words?

We usually think of potential energy as being a measureable entity stored in a material existence, but perhaps that is not the only reality. Perhaps something as intangible as words is the linkage "Between the idea / And the reality." Think of the potential energy within the words:

"When in the course of human events ..."
- Thomas Jefferson -

"I have a dream ..."
- Martin Luther King Jr. -

"... blood, sweat, and tears."
- Winston Churchill -

Perhaps the despair of "The Hollow Men" comes from the actuality of entropy and the fact that energy is running downhill. They have consumed themselves. Their inevitable conclusion comes from inaction, from neglect, from exhaustion, from ignorance of their own potentials. We consider the possibility when we reflect upon the last two lines:

This is the way the world ends
Not with a bang but a whimper.

CHAPTER IV CHEMISTRY

> Thought often comes clad in the strangest clothing:
> So Kekule the chemist watched the weird route
> Of eager atom-serpents writhing in and out
> And waltzing tail to mouth. In that absurd guise
> Appeared benzene and analine, their drugs and
> their dyes.
> -Robert Ranke Graves-

Alchemy to Biochemistry

> So the learn'd Alchemist exulting sees
> Rise in his bright mattrass Diana's trees;
> Drop after drop, with just delay he pours
> The red-fumed acid on Potosi's ores;
> With sudden flash the fierce bullitions rise,
> And wide in air the gas phlogistic flies;
> Slow shoot, at length, in many a brilliant mass
> Metallic roots across the netted glass.

> Branch after branch extend their silver stems,
> Bud into gold, blossom into gems.
> - from "Botanic Garden" -
> - Erasmus Darwin -

From the beginning, probably no other science has been so naturally based on mysticism and magic as has been chemistry. Outside a domain of practical applications which actually preceeded recorded history -- such as the preparation of wine and vinegar, pottery making, elementary metallurgy, and dyeing -- chemistry was perceived to be the magical key in man's persistent quest for the elusive goals of good health, longevity, immortality, and gold. Indian Vedic texts a thousand years before the Christian era stressed the linkage between gold and immortality: "Gold is indeed fire, light, and immortality."[1] Farther east the Chinese also linked gold to good health. The aristocracy used gold plates to eat upon, not so much because gold represented wealth, but because they believed gold contributed to good health. The allure of gold has possibly been described best by Thomas Hood (1799-1845) in his poem by that name.

> Gold! Gold! Gold! Gold!
> Bright and yellow, hard and cold,
> Molten, graven, hammered, and rolled,
> Heavy to get and light to hold;
> Hoarded, bartered, bought and sold,
> Stolen, borrowed, squandered, doled;
> Spurned by the young, but hugged by the old
> To the very verge of the churchyard mold;
> Price of many a crime untold.
> - "Gold!" -
> - Thomas Hood -

We should not find it surprising then that the consuming passion of the ancient chemist, the alchemist, was the pursuit of the process to change base elements into gold. With health, wealth, and immortality as the envisioned outcome, alchemy survived well into the mid 1600s.

The fervor of the alchemist's drive to create gold was simultaneously inflamed and subdued by Paracelsus (1493-1541), one of the most enigmatic and "bombastic" persons in the history of science and about whom we shall refer again. He was the uncrowned prince among the alchemists of the day. He was a prolific writer of over 230 publications dealing not only with alchemy, but also medicine, astrology, magic, and theology. An excerpt from one of his writings tells his fellow alchemists of the futility of the quest for gold, yet he dangles before them the one possibility for success, but then withdraws the hope.

> ... you should know that alchemy is nothing but the art which makes the impure into the pure through fire ... It can separate the useful from the useless, and transmute it into its final substance and its ultimate essence. The transmutation of metals is a great mystery of nature. However laborious and difficult this task may be, whatever impediments and obstacles may lie in the way of its accomplishment, this transmutation does not go counter to nature, nor is it incompatible with the order of God, as is falsely asserted by many persons. But the base, impure five metals -- that is, copper, tin, lead, iron, and quicksilver -- cannot be transmuted into the nobler, pure, and perfect metals -- namely, into gold and silver -- without a *tinctura*, or without the philosopher's stone.[2]
>
> *   *   *   *   *
>
> Here on earth the celestial fire is cold, rigid, and frozen fire. And this fire is the body of gold. Therefore all we can do with it by means of our fire is to dissolve it and make it fluid, just as the sun thaws snow and ice and makes them liquid. In other words, fire has not the power to burn fire, for gold itself is nothing but fire. In heaven it is dissolved, but on earth it is solidified ... But

to write more about this mystery is forbidden and further revelation is the prerogative of the divine power. For this art is truly a gift of God. Wherefore not everyone can understand it. For this reason God bestows it upon whom He pleases, and it cannot be wrested away from Him by force; for it is His will that alone shall be honored in it and that through it His name be praised for ever and ever.[3]

The poetic play, "The Alchemist," written by poet laureate Ben Johnson in 1610, was a highly entertaining and dramatic satire on human greed. In it, he captured well the excitement and dreams of one under the spell of the alchemist. The sought for Philosopher's Stone would unlock all of the secrets of metallurgy enabling the instantaneous conversion of all possessions into gold; the Elixir of Life would give eternal youth, vigor, and immortality. We may also find it interesting that not only was there a growing awareness of the "pseudo-scientific" nature of the alchemist-confidence man, but also that certain health hazards of the occupation were known. There is little doubt that the work with lead, mercury, and other heavy metals caused physiological damage. The character, Mammon, exhorted that he will be able to use the Elixir to "... repair this braine, / Hurt wi' the fume o' the metals."

MAMMON. .... This night, I'll change
All, that is mettal, in thy house, to gold.
And, early in the morning, will I send
To all the plumbers, and the pewterers,
And buy their tin, and lead up: and to Lothbury,
For all the copper.
SURLY. What, and turne that too?
MAMMON. Yes, and I'll purchase Devonshire, and
Cornwaile, And make them perfect Indies! You
Admire now?
SURLY. No, faith.

MAMMON. But when you see th' effects of
 The great med'cine of which one part projected
 On a hundred of Mercurie, or Venus, or the
 Moone, shall turne it, to as many of the Sunne;
 Nay, to a thousand, so ad infinitum:
 You will beleeve me.
SURLY. Yes, when I see't, I will.
 But, if my eyes doe cossen me so (and I
 Giving 'hem no occasion) sure, I'll have
 A whore, shall pisse 'hem out, next day.
MAMMON. Ha! Why?
 Doe you thinke, I fable with you? I assure you,
 He that has once the flower of the sunne,
 The perfect ruby, which we call elixir,
 Not onely can doe that, but by it's vertue,
 Can confer honour, love, respect, long life,
 Give safetie, valour: yea, and victorie,
 To whom he will. In eight and twentie dayes,
 I'll make an old man, of fourescore, a childe.
SURLY. No doubt, hee's that alreadie.
MAMMON. Nay, I meane,
 Restore his yeeres, renew him, like an eagle,
 To the fifth age; make him get sonnes, and daughters,
 Young giants; as our Philosophers have done
 (The antient Patriarkes afore the floud)
 But taking, once a weeke, on a knives point,
 The quantitie of a graine of mustard, of it:
 Become stout Marses, and beget young Cupids.
SURLY. The decay'd Vestall's of Pickt-hatch would thanke
 You. That keepe the fire a-live, there.
MAMMON. 'Tis the secret
 Of nature, naturiz'd 'gainst all infections,
 Cures all diseases, comming of all causes,
 A month's griefe, in a day; a yeeres, in twelve:
 And, of what age soever, in a month.
 Past all the doses, of your drugging Doctors.

>   I'll undertake, withall, to fright the plague
>   Out o' the kingdome, in three months.
> SURLY.   And I'll
>   Be bound, the players shall sing your praises, then
>   Without their poets.
> MAMMON.   Sir, I'll doo't. Meanetime,
>   I'll give away so much, unto my man,
>   Shall serve th' whole citie, with preservative,
>   Weekely, each house his dose, and at the rate---
> SURLY. As he built the water-worke, do's with water?
> MAMMON. You are incredulous.
> SURLY.   Faith, I have a humor,
>   I would not willingly be gull'd. Your stone
>   Cannot transmute me.
> MAMMON.   Pertinax, my Surly,
>   Will you beleeve antiquitie? recordes?
>   I'll shew you a booke, where Moses, and his sister,
>   And Salomon have written, of the art;
>   Ay, and a treatise penn'd by Adam.
> SURLY.   How!
> MAMMON. O' the Philosophers stone, and in high-Dutch.
> SURLY. Did Adam write, sir, in high-Dutch?
> MAMMON.   He did:
>   Which proves it was the primitive tongue.
> SURLY.   What paper?
> MAMMON. On cedar board.
> SURLY.   O that, indeed (they say)
>   Will last 'gainst wormes.
> MAMMON.   'Tis like your Irish wood,
>   'Gainst cob-webs. I have a peece of Jasons fleece, too,
>   Which was no other, then a booke of alchemie,
>   Writ in large sheepe-skin, a good fat ram-vellam.
>   Such was Pythagoras' thigh, Pandora's tub;
>   And, all that fable of Medeas charmes,
>   The manner of our worke: The Bulls, our fornace,
>   Still breathing fire; our argent-vive, the Dragon:

The Dragons teeth, mercury sublimate,
That keepes the whitenesse, hardnesse, and the biting;
And they are gather'd, into Jason's helme,
(Th' alembeke) and then sow'd in Mars his field,
And, thence, sublim'd so often, till they are fix'd.
Both this, th' Hesperian garden, Cadmus storie,
Jove's shower, the boone of Midas, Argus eyes,
Boccace his Demogorgon, thousands more,
All abstract riddles of our stone.

[Enter Face.]

How now?
Doe wee succeed? Is our day come? and hold's it?
FACE. The evening will set red, upon you, sir;
You have colour for it, crimson: the red ferment
Has done his office. Three houres hence, prepare you
To see projection.
MAMMON.   Pertinax, my Surly,
Againe, I say to thee, aloud: be rich.
This day, thou shalt have ingots: and, to morrow,
Give lords th' affront. Is it, my Zephyrus, right?
Blushes the bolts-head?
FACE.   Like a wench with child, sir,
That were, but now, discover'd to her master.
MAMMON. Excellent wittie Lungs! My onely care is,
Where to get stuffe, inough now, to project on,
This towne will not halfe serve me.
FACE.   No, sir? Buy
The covering off o' churches.
MAMMON.   That's true.
FACE.   Yes.
Let 'hem stand bare, as doe their auditorie.
Or cap 'hem, new, with shingles.
MAMMON.   No, good thatch:
Thatch will lie light upo' the rafters, Lungs.
Lungs, I will manumit thee, from the fornace;

I will restore thee thy complexion, Puffe,
Lost in the embers; and repaire this braine,
Hurt wi' the fume o' the mettalls.
FACE. I have blowne, sir,
Hard, for your worship; throwne by many a coale,
When 'twas not beech; weigh'd those I put in, just,
To keepe your heat, still even; These bleard-eyes
Have wak'd, to reade your severall colours, sir,
Of the pale citron, the greene lyon, the crow,
The peacocks taile, the plumed swan.
MAMMON. And, lastly,
Thou hast descryed the flower, the sanguis agni?
FACE. Yes, sir.
MAMMON. Where's master?
FACE. At's praisers, sir, he,
Good man, hee's doing his devotions,
For his successe.
MAMMON. Lungs, I will set a period,
To all thy labours: Thou shalt be the master
Of my seraglia.
FACE. Good, sir.
MAMMON. But doe you heare?
I'll geld you, Lungs.
FACE. Yes, sir.
MAMMON. For I doe meane
To have a list of wives, and concubines,
Equall with Salomon; who had the stone
Alike, with me: and I will make me, a back
With the elixir, that shall be as tough
As Hercules, to encounter fiftie a night.
Th'art, thou saw'st it bloud?
FACE. Both bloud, and spirit, sir.
MAMMON. I will have all my beds, blowne up; not stuft:
Downe is too hard. And then, mine oval roome,
Fill'd with such pictures, as Tiberius tooke
From Elephantis: and dull Aretine

But coldly imitated. Then, my glasses,
Cut in more subtill angles, to disperse
And multiply the figures, as I walke
Naked between my succuba. My mists
I'le have of perfume, vapor'd 'bout the roome,
To lose our selves in; and my baths, like pits
To fall into: from whence, we will come forth,
And rowle us drie in gossamour, and roses.
(Is it arriv'd at ruby?) -- Where I spie
A wealthy citizen, or rich lawyer,
Have a sublim'd pure wife, unto that fellow
I'll send a thousand pound, to be my cuckold.
FACE. And I shall carry it?
MAMMON. No. I'll ha' no bawds,
But fathers, and mothers. They will doe it best.
Best of all others. And, my flatterers
Shall be the pure, and gravest of Divines,
That I can get for money. My mere fooles,
Eloquent burgesses, and then my poets
The same that writ so subtly of the fart,
Whom I will entertaine, still, for that subject.
The few, that would give out themselves, to be
Court and towne-stallions, and, each where, belye
Ladies, who are knowne most innocent, for them;
Those will I begge, to make me eunuchs of:
And they shall fan me with ten estrich tailes
A piece, made in a plume, to gather wind.
We will be brave, Puffe, now we ha' the med'cine.
My meat, shall all come in, in Indian shells,
Dishes of agate, set in gold, and studded
With emeralds, saphyres, hiacynths, and rubies.
The tongues of carpes, dormice, and camels heeles,
Boil'd i' the spirit of Sol, and dissolv'd pearle,
(Apicius diet, 'gainst the epilepsie)
And I will eat these broths, with spoones of amber,
Headed with diamant, and carbuncle.

> My foot-boy shall eate phesants, calverd salmons,
> Knots, godwits, lamprey's: I my selfe will have
> The beards of barbels, serv'd, in stead of sallades;
> Oild mushrooms: and the swelling unctuous paps
> Of a fat pregnant sow, newly cut off,
> Drest with an exquisite, and poynant sauce;
> For which, Ile say unto my cooke: There's gold,
> Goe forth, and be a knight.
> FACE.   Sir, I'll goe looke
> A little, how it heightens.
> MAMMON.   Doe. My shirts
> I'll have of taffata-sarsnet, soft, and light
> As cobwebs; and for all my other rayment
> It shall be such, as might provoke the Persian,
> Were he to teach the world riot anew.
> My gloves of fishes, and birds-skins, perfum'd
> With gummes of paradise, and easterne aire ----
> SURLY. And do you thinke, to have the stone, with this?
> MAMMON. No, I doe thinke, t'have all this, with the stone.
>   - from *The Alchemist* Act II, sc.1,2 -
>   - Ben Johnson -

Paracelsus (Philippus Aureolus Theophrastus Bombastus von Helmholtz), alchemist and physician of the sixteenth century, was as outspoken as he was brilliant: "My proofs derive from experience to my own reasoning, and not from reference to authorities .... It seems imperative to bring medicine back to its original laudable state, and, aside from striving to cleanse it of the dregs left by barbarians, to purify it of the most serious errors."[4]

He was essentially the first to identify life as a chemical process. His discovery of the relationship between goiter in the parent and cretinism in the child led to his conclusion that illness was due to imbalances in the correct chemical ingredients. Although he oversimplified the range of chemical treatments to that of mercury, sulfur and salt, his theory was a large step from the imbalance of humours. His treatments, which foreshadowed modern medical practices, compelled physicians to learn a little about chemistry. "How can I praise those who are physicians and not alchemists at the same time?" he said. "For never must knowledge and preparation, that is to say, medicine and alchemy, be separated from each other."[5] As for his emphasis on mercury, sulfur, and salts -- the use of mercury chloride (calomel), sulfur (molassess and sulfur), and salts (Epsom salts) has not totally passed into obscurity.

Sweeping changes continued to occur in science during the eighteenth and nineteenth centuries. The discovery of oxygen eliminated the idea of phlogiston; Dalton's Atomic Theory eliminated the idea of the continuous nature of primary material; the synthesis of urea eliminated the idea of the "vital force"; the discovery of radioactivity eliminated the idea of the indestructable atom. And the twentieth century has continued with each new discovery on the heels of another -- nylon, synthetic vitamins, synthetic rubber, wonder drugs, RNA/DNA. Chemistry has come unto its own as one of the important influences on modern man. Who can argue with the advertising slogan, "Better living through chemistry"? In the poem by Hugh MacDiarmid (1892-1978) having the unusual title, "The Kind of Poetry I Want," we find an expression of the beneficial role that some poets believe chemistry plays in contemporary life.

> And, constantly, I seek
> A poetry of facts. Even as
> The profound kinship of all living substance
> Is made clear by the chemical route.
> Without some chemistry one is bound to remain
> Forever a dumbfounded savage
> In the face of vital reactions.
> The beautiful relations
> Shown only by biochemistry
> Replace a stupefied sense of wonder
> With something more wonderful
> Because natural and understandable.

Nature is more wonderful
When it is at least partly understood.
Such an understanding dawns
On the lay reader when he becomes
Acquainted with the biochemistry of the glands
In their relation to diseases such as goiter
And in their effects on growth, sex, and reproduction.
He will begin to comprehend a little
The subtlety and beauty of the action
Of enzymes, viruses, and bacteriophages,
Those substances which are on the borderland
Between the living and the non-living
He will understand why the biochemist
Can speculate on the possibility
Of the synthesis of life without feeling
That thereby he is shallow or blasphemous.
He will understand that, on the contrary,
He finds all the more
Because he seeks for the endless
--"Even our deepest emotions
May be contradicted by traces
Of a derivative of phenanthrene!"
    - from "The Kind of Poetry I Want" -
    - Hugh MacDiarmid -

But the chemist and the chemical industry are not universally perceived as the altruistic benefactor for twentieth century man. There is also a vision of a hard, dispassionate manipulator of the inanimate whose concern lies only with the attractions between cold, lifeless atoms. The chemist is viewed by his detractors as one who can see water in no terms other than H_2O whether it is a waterfall, a flood, a crystalline pool, or a tear; as one who has graced man's life with dynamite, DDT, and agent orange; as well as one who represents synthetic vanillin as the real thing because its chemical composition is identical to vanilla pressed from the seed.

The poets-of-distress, who find reasons to criticize chemistry's effects on life and modern man, frame their messages in diverse ways. In the style of open poetry we find "Ice Cream Cone" to be a silent satire. It is silent because the words themselves say nothing of an emotional nature and, although seen everyday on labels and advertisements, are ignored or taken for granted. But the same words, within the context of the poem satirically imply the deceit and contamination in the world today as opposed to the idealized purity of the past. The unaware reader is brusquely reminded that he lives on the edge of an artificial reality.

> Flour, cereal, sugar, starch,
> vegetable shorting,
> salt, protein, gum leavening,
> propylene glycol.
> Certified colors, artificial flavors.
> Chocolate coating containing
> cocoa, vegetable oil (containing
> an emulsifier and tenox 2
> less than .05%).
> Tenox 2
> is an antioxidant containing
> butylated hydroxyanisole,
> propyl gallate, citric acid,
> propylene glycol.
>   - "Ice Cream Cone" -
>   - Ronald Gross -

And yet there exists among some poets the desire for the control and mastery exhibited by the chemist. Nobel poet Sully-Prudhomme's (1839-1907) poem, "The Naked World," describes the chemist at work fathoming, imposing, controlling, and directing the "secret affinities" of the substances with which he works. We can read into his words an admiration and a longing for this capacity. If we were to end our reading before the last three lines, we might conclude that the poem was simply a tribute to the chemist. But the poem says more than that. It ends with a plea that the writer could possess the same kinds of insights into man, and that he could possess the same kinds of capacities for understanding and modeling man that the chemist has for his atoms and molecules.

> Surrounded by beakers, by strange coils,
> By ovens and flasks with twisted necks,
> The chemist, fathoming the whims of attractions,
> Artfully imposes on them their precise meetings.
>
> He controls their loves, hidden until now,
> Discovers and directs their secret affinities,
> Unites them and brings about their abrupt divorces,
> And purposefully guides their blind destinies.
>
> Teach me then, to read right to the bottom of your alembics,
> O sage, who understands these stark forces,
> And the inside of the world beyond all color.
>
> Lead me, I pray, into this dark kingdom:
> It is toward inward realities that I strive;
> Outward forms, too beautiful, beget only sorrows.
> - "The Naked World" -
> - Sully-Prudhomme -

The Elements

You will recall from the chapter dealing with atoms that two major questions dominated ancient natural philosophy. One was whether matter was continuous as the Aristotlean view described, or particulate as Democritus theorized? The viewpoint that matter existed as discrete particles which we now know as atoms could not overpower the argument of Aristotle's continuous material. The continuous theory was much more logical when we consider the prevailing belief about primary material. The four elements were Air, Earth, Fire, and Water; three of them were obviously not particulate.

It was during this same general period that the Hippocratic treatise, "On the Nature of Man," identified four humours as primary material. A preponderance of each was associated with the four ages of man, the four seasons, and the four major organs -- heart, brain, liver, and spleen. Imbalances from the desired normalcy caused illness and death. Plato, futhermore, associated psychological behavior to the humours and described them in terms of melancholic, sanguine, choleric, and phlegmatic. Excess phlegm in the brain caused epilepsy; excess bile caused frenzy. Incredibly the existence of humours to explain illness and emotion persisted well into the eighteenth century. Physicians, poets, and writers, used the humours to explain behavior. The following lines were not atypical.

> His life was gentle, and the elements
> So mixed in him that Nature might stand up
> And say to all the world, "This was a man."
> - from *Julius Caesar* -
> - William Shakespeare -

> The elements, that do man's house compose
>   Are all his chiefest foes;
> Fire, air, earth, water all are at debate,
>   Which shall predominate.
> - from "On the Death of My Dear
>   Brother" -
> - William Hammond -

Christopher Marlowe, writing at about the same period, expressed similar notions about the innate influences of humours on man. Today we scoff at this primitive belief and discount the idea that Elements and Humours exist within man that determine his behaviors and capacities. Yet, if we replace Marlowe's words "four elements" with the contemporary language of deoxyriboneucleic acid (DNA), we are faced with a new reality. Biochemical studies provide evidence that our obvious hereditary characteristics -- size, shape, color -- are determined by DNA. And more evidence points to a conclusion that talent, emotion, and personality may well be determined by that same wondrous molecule. Thus, the irony continues. The *four* Elements, the *four* Humours -- mystical and magic for over two thousand years -- have been replaced today by DNA which is composed of *four* smaller molecules: guanine, cytosine, thymine, and adenine. The proper balance is required for health and life. Does that sound familiar?

>Nature that framed us of our four elements,
>Warring within our breasts for regiment,
>Doth teach us all to have aspiring minds:
>Our souls, whose faculties can comprehend
>The wondrous architecture of the world
>And measure every wandering planet's course,
>Still climbing after knowledge infinite,
>And always moving as the restless spheres,
>Will us to wear ourselves, and never rest . . .
>   - from *Tamburlaine* Act II, sc VII -
>   - Christopher Marlowe -

The seventeenth century was a period of disputation, controversy, and debate as the new ideas and theories of science ran head-on into the older natural philosophy. The new science was having an unsettling effect on groups of intellectuals who were bound to the older philosophy. Characteristic of the poetry of this group was the piece written by John Donne (1573-1631) entitled "The First Anniversary." His preface to the poem gives us an added insight into his personal feelings, and sets the stage for the section that follows.

An Anatomy of the World
Wherein, By the occasion of the untimely Death of
Mistress Elizabeth Drury the frailty and decay of this
whole World is represented.
>- Preface to "The First Anniversary" -
>- John Donne -

Donne definitely viewed the new era as an indication of a severe deterioration in the natural universe and a disruptive fragmentation of the order -- "all coherence gone."

>And new Philosophy calls all in doubt,
>The Element of fire is quite put out;
>The Sun is lost, and th' earth, and no man's wit
>Can well direct him where to looke for it.
>And freely men confesse that this world's spent,
>When in the Planets, and the Firmament
>They seeke so many new; they see that this
>Is crumbled out againe to his Atomies.
>'Tis all in peeces, all cohaerence gone;
>All just supply, and all Relation:
>Prince, Subject, Father, Sonne, are things forgot,
>For every man alone thinkes he hath got
>To be a Phoenix, and that there can bee
>None of that kinde, of which he is, but hee. . . .
>- from "The First Anniversary" -
>- John Donne -

The Subtleties

On occasions poets use the characteristics of the chemical elements in a more subtle approach to their messages. Typical are some of the poems that deal with the concepts of value and duration. Robert Frost used a unique approach as he expressed the concept of duration -- permanence and impermanence -- in his poem, "Nothing Gold Can Stay."

> Nature's first green is gold
> Her hardest hue to hold.
> Her early leaf's a flower;
> But only so an hour.
> Then leaf subsides to leaf.
> So Eden sank to grief,
> So dawn goes down to day.
> Nothing gold can stay.
> -"Nothing Gold Can Stay"-
> -Robert Frost-

    From the title, our first reaction is that Frost is wrong. We all know that gold is permanent; it does not rust, does not decay, does not dissolve. But then, as we read the poem, we recognize that he is referring to the golden leaf, the golden flower, and the golden sunrise. True enough, this gold does not stay. And, therein, lies the craft of Frost. He makes us believe from the title that he is speaking about gold metal, and we take exception with him. But he speaks not of the metal but of the pigment and light; now we agree. Yet in our agreement we reexamine the title and a realization evolves that he was correct after all--*nothing* gold can really stay.

    Duration is an important but confusing component of man's experience, and by duration we mean the impermanent-permanent dichotomy. Philosophically, if not practically, how can these opposites exist simultaneously?

    The ancient Chinese writings tell us that relative experiences such as good and evil, light and darkness, hot and cold, are not truly opposed but are all complementary in that each depends upon the other for its existence. The scientist and philosopher, E. R. Scerri, writes that "the juxtaposition of opposite qualities is seen to give rise to diversity in creation, and the wise person is he or she who can reconcile the apparent conflict of opposites both within his or her own being and the external world."[6] In the *Tao-*

te ching is written:

> Under heaven all can see beauty as beauty only
> because there is ugliness.
> All can know good only because there is evil.
> Therefore having and not having arise together.
> Difficult and easy complement each other.
> High and low rest on each other.
> Voice and sound harmonise each other.
> Front and back follow one another.
> Therefore the sage goes about doing nothing,
> teaching not talking.[7]
> -from *Tao-te ching*, ch. 2-
> -Lao Tzu-

W. A. Watts, writing in "Psychotherapy: East and West" alleges that it is difficult for westerners to assimilate fully the notion that polar opposites are the extremes of a single whole concept.[8] But those in the East, especially the Taoist poets, see the opposites as opposites, but do not recognize them as dichotomies.

> In order to contract,
> It is necessary first to expand.
> In order to weaken,
> It is necessary first to strengthen.
> In order to destroy,
> It is necessary first to promote.
> In order to grasp,
> It is necessary first to give.
> This is called subtle light.
> The weak and the tender overcome the hard and
> the strong.[9]
> -from *Tao-te ching*, ch. 36-
> -Lao Tzu-

The chemist, along with the Taoist, is comfortable with

contradictions that complement one another. The atom is composed of positive and negative components; positive ions bond to negative ions; acids and bases neutralize one another; and oxidation and reduction must occur simultaneously. The Chinese derive a sense of permanence from the polar opposites and their transitory states.

To a degree, Wallace Stevens (1879-1955) built upon this idea of contradictions and impermanence in his poem entitled "The Glass of Water." He begins with the observation that water and glass are of one kind although momentarily appearing to be different. Contemporary critics see this as an expression that existence is simply a state of being, and that man's concepts of reality are relative and of necessity existential.[10]

> That the glass would melt in heat,
> That the water would freeze in cold,
> Shows that this object is merely a state,
> One of many, between two poles. So,
> In the metaphysical, there are these poles.
>
> Here in the centre stands the glass. Light
> Is the lion that comes down to drink. There
> And in that state, the glass is a pool.
> Ruddy are his eyes and ruddy are his claws
> When light comes down to wet his frothy jaws
>
> And in the water winding weeds move round.
> And there and in another state -- the refractions,
> The *metaphysica,* the plastic parts of poems
> Crash in the mind -- But, fat Jocundus, worrying
> About what stands here in the center, not in the glass,
>
> But in the centre of our lives, this time, this day,
> It is a state, this spring among the politicians
> Playing cards. In the village of the indigenes,

One would have still to discover. Among the dogs and dung,
One would continue to contend with one's ideas.
    -"The Glass of Water"-
    -Wallace Stevens-

    John Vance Cheney's (1848-1922) poem, "The Happiest Heart," is an interesting approach to the concept of value and of duration. He defines fame as an example of value, and correlates it quite naturally with duration -- the permanence and impermanence of fame -- and states firmly that the simpler things of life have a more lasting value.

    The key to Cheney's message is the parallel drawn between rust and fame. We can easily visualize the bright, shining sword of fame, and we have all had the experience of seeing the decay and disintegration of rusting iron. The statement that "... the rust will find the sword of fame," warns of the temporary nature of fame.

    The chemist, however, as opposed to the poet, might see another sword of fame. His could be chromium plated whereas the "rust" of chromium is a thin, transparent layer which actually protects the metal from further corrosion and degradation. Yet he would be the first to admit that the corrosion is not eliminated, it is just delayed. Or he may evision a graphite sword whose rust is not spontaneous but, when intense heat is applied, becomes gaseous carbon dioxide so that the graphite sword could vanish without a trace. Or he may visualize a shiny radium metal sword that slowly but surely decays to gray, soft lead. Neither sword nor fame is so lasting that "Time will not tear it down."

> Who drives the horses of the sun
> Shall lord it but a day;
> Better the lowly deed were done,
> And kept the humble way.
>
> The rust will find the sword of fame,
> The dust will hide the crown;

> Ay, none shall nail so high his name
> Time will not tear it down.
>
> The happiest heart that ever beat
> Was in some quiet breast
> That found the common daylight sweet,
> And left to Heaven the rest.
>   - "The Happiest Heart" -
>   - John Vance Cheney -

Inevitably, the element gold is associated with value and worth. The linkage is obvious. It is therefore obvious that the cautions and warnings about false values would be based to some extent upon gold. In Shakespeare's, *Merchant of Venice*, he used three common elements and assigned to each an illusive characteristic. In Act II, scene 7, we read the following dialog between Portia and the Prince of Morocco:

> Portia. Go, draw aside the curtains, and discover
> The several caskets to this noble prince.
> Now make your choice.
>
> Prince. The first, of gold, which this inscription bears:
> Who chooseth me shall gain what many men desire.
> The second, silver, which this promise carries:
> Who chooseth me shall get as much as he deserves.
> This third, dull lead, with warning all as blunt:
> Who chooseth me must give and hazard all he hath.
> How shall I know if I do choose the right?
>
> Portia. The one of them contains my picture, prince:
> If you choose that, then I am yours withal.

After much anxious deliberation the Prince selects one of the caskets-- the gold one -- and naturally it is the wrong choice. In place of the treasure he sought, he found instead a scroll.

                [He unlocks the golden casket.]
Prince.   O hell! what have we here?
A carrion Death, within whose empty eye
There is a written scroll. I'll read the writing.
            'All that glistens is not gold;
            Often have you heard that told:
            Many a man his life hath sold
            But my outside to behold:
            Gilded tombs do worms infold.
            Had you been as wise as bold,
            Young in limbs, in judgement old,
            Your answer had not been inscroll'd:
            Fare you well; your suit is cold.'
    Cold, indeed; and labour lost:
    Then, farewell, heat, and welcome, frost!
Portia, adieu. I have too griev'd a heart
To take a tedious leave: thus losers part.
            - from *Merchant of Venice* -
            - William Shakespeare -

Where then is the real treasure one seeks? In "The Complaint; Or Night Thoughts on Life, Death and Immortality," the author, Edward Young (1683-1765), gives clear directions to the reader: "... seek it in thyself."

Where, thy true treasure? Gold says, 'Not in me:'
And, 'Not in me,' the diamond. Gold is poor;
India's insolvent: Seek it in thyself,
Seek in thy naked self, and find it there;
In being so descended, formed, endowed;
Sky-born, sky-guided, sky-returning race!
Erect, immortal, rational, divine!
In senses, which inherit earth and heavens,
Enjoy the various riches nature yields;
Far nobler! give the riches they enjoy;

> Give taste to fruits and harmony to groves;
> Their radiant beams to gold, and gold's bright fire;
> Take in, at once, the landscape of the world,
> At a small inlet, which a grain might close,
> And half create the wondrous world they see.
> Our senses, as our reason, are divine.
> But for the magic organ's powerful charm,
> Earth were a rude, uncoloured chaos still.
> Objects are but the occasion; ours the exploit;
> Ours is the cloth, the pencil, and the paint
> Which nature's admirable picture draws;
> And beautifies creation's ample dome.
> Like Milton's Eve, when gazing on the lake,
> Man makes the matchless image man admires:
> Say then, shall man his thoughts all sent abroad,
> Superior wonders in himself forgot,
> His admiration waste on objects round,
> When heaven makes him the soul of all he sees?
> Absurd; not rare! so great, so mean, is man.
> Ours is the cloth, the pencil, and the paint
> Which nature's admirables picture draws;
> And beautifies creation's ample dome.
> Like Milton's Eve, when gazing on the lake,
>
>   - from "The Complaint: Or Night
>   Thoughts on Life, Death, and
>   Immortality" -
>   - Edward Young -

Value is also the message in the complicated poem entitled "Definitions for Mendy," written by the contemporary poet, David Antin. He speaks of value in terms of relativity -- "Thirst is a desert / Value a glass of water." He speaks of value in terms of contradictions -- what seems to be may not be -- "the presence of the dead is imaginary / the absence is real."

We must also consider his observation about radium: "radium is a value that is always declining / radium is a value that is always disappearing." Here he says that value and worth spontaneously decline just as the quantity of radium declines and disappears through its natural process of radioactivity and transmutation. Ironically, it is the radiation feature that produces the value; it is that same radioactive factor that leads to the reduction of the value. Just as ripening gives value to the fruit, ripening also leads to the destruction of the fruit and the elimination of its value.

His parallel with the metal lead makes an interesting comparison. "Lead is also a value / but it is less bright than radium." Lead does not have the initial value of radium, but lead endures. The last four lines of the portion of the poem that have been quoted here remind us of the variety of factors that contribute to value: strength, power, brightness, and duration.

> loss is an unintentional decline in our disappearance of
>   a value arising from a contingency
> a value is an efficacy a power a brightness
> it is also a duration
>
> to lose something keys hair someone
> we suffer at thought
> he has become absent imaginary false
> a false key will not turn a true lock
> false hair will not turn grey
> mendy will not come back
> but longing is not imaginary
> we must go down into ourselves
> down to the floor   that is not imaginary
> where hunger lives and thirst
> hunger imagine bread   thirst imagine water
> the glass of water slips to the floor
> thirst is a desert
> value a glass of water
> loss is the glass of water slipping to the floor
> loss is the unintentional decline in or disappearance

      of a glass of water arising from a contingency
the glass   pieces of glass
the floor is a contingency
the floor is a floor
is a contingency
made of wood
the fire is a contingency
the bread is burned
burning is not a contingency
the presence of the dead is imaginary
the absence is real
henceforth it will be his manner of appearing
so he appears   in an orange jacket and workpants
and a blue demin shirt
his hair is black his eyes are black
and a blue crab is biting his long fingers
he is trying to hold the bread
he is trying to bring the water to his mouth
his mouth is a desert
the glass of water will not come
the glass of water keeps slipping through his fingers
the floor is made of wood it is burning
it is covered with pieces of glass
arising from a contingency
his face is the darkened face of a clock
it is marked with radium
the glass is falling from his face
the face of a clock in which there is a salamander
whose eyes are bright with radium
radium is a value that is always declining
radium is a value that is always disappearing
lead is also a value
but it is less bright than radium

loss is an unintentional decline in or disappearance of
  a value rising from a contingency

a value is an efficacy a power a brightness
it is also a duration
            - from "Definitions for Mendy" -
            - David Antin -

Analogies

Each substance has certain objective properties or characteristics that set it apart from the other substances . . . properties such as color, natural state, density, solubility and a variety of others. In addition to these tangible and to a large degree measureable characteristics, many have an emotional or subjective quality associated with them. Poets have quite naturally used the qualities of these substances to draw parallels with human existence and experiences.

Emily Dickinson (1830-1886), the shy New England poet, provides an example of the creation of a mood by describing it in terms of an element.

    After great pain a formal feeling comes--
    The nerves sit ceremonious like tombs;
    The stiff heart questions--was it He that bore?
    And yesterday--or centuries before?

    The feet mechanical go round
    A wooden way,
    Of ground or air of Ought,
    Regardless grown;
    A quartz contentment like a stone.

    This is the hour of lead
    Remembered if outlived
    As freezing persons recollect
    The snow--

> First chill, then stupor, then
> The letting go.
>> - "After Great Pain" -
>> - Emily Dickinson -

We think of lead as dull, gray, heavy, inert, poisonous, and deadeningly quiet. It is the inert (dead) end of product of the spontaneous energy emission (life) of radioactive elements. We can feel ourselves being weighted down by the dense, drab mass. The whole connotation is that of weighty lifelessness. The mood of gloom permeates completely.

We can find the diametrically opposite mood created in Frost's poem, "Innate Helium." He uses the bouyancy of gas to symbolize the uplift of one's spirit through religious faith. The reader can almost feel an expansion and lightness created as though he were a balloon filled with helium -- tugging at some restraint, escaping from some unwanted grasp. This uplifting force, this uplifting faith, "must be innate."

> Religious faith is a most filling vapor.
> It swirls occluded in us under tight
> Compression to uplift us out of weight--
> As in those buoyant bird bones thin as paper,
> To give them still more buoyancy in flight.
> Some gas like helium must be innate.
>> - "Innate Helium" -
>> - Robert Frost -

A more subtle use of chemical properties is demonstrated by poet and artist, Washington Allston (1779-1843). The particular line, "To plough up light that ever round us streamed," has the reference in chemistry to the luminous phosphors that are present in sea water and which sparkle and glow as silver streaks when the dark waters of night are disturbed. Allston equates this phenomenon to Coleridge's existence. Coleridge was a shinning shimmering radiance in an otherwise dark ocean of ignorance and insensitivity. As the luminous phosphors light up the darkness, so shall his truths illuminate the soul.

> And thou art gone, most loved, most honored friend!
> No, nevermore thy gentle voice shall blend
> With air of Earth its pure ideal tones,
> Binding in one, as with harmonious zones,
> The heart and intellect. And I no more
> Shall with thee gaze on that unfathomed deep,
> The Human Soul,-- as when, pushed off the shore,
> Thy mystic bark would through the darkness sweep,
> Itself the while so bright! For oft we seemed
> As on some starless sea, -- all dark above,
> All dark below, -- yet, onward as we drove,
> To plough up light that ever round us streamed.
> But he who mourns is not as one bereft
> Of all he loved: thy living Truths are left.
> - "On the Late S. T. Coleridge" -
> - Washington Allston -

Thomas Stanley in a moral poem written in 1647 used among other things common iron to create an analogy with love.

> Ask the Empress of the night
>   How the hand which guides her sphear,
> Constant in unconstant light,
>   Taught the waves her yoke to bear,
> And did thus by loving force
> Curb or tame the rude seas course.
>
> Ask the female Palme how shee
>   First did woo her husbands love;
> And the Magnet, ask how he
>   Doth th'obsequious iron move;
> Waters, plants and stones know this,

That they love, not what love is.

> Be not then less kind that these,
>   Or from love exempt alone,
> Let us twine like amorous trees,
>   And like rivers melt in one;
> Or if thou more cruell prove
> Learne of steel and stones to love.
>   - "The Magnet" -
>   - Thomas Stanley -

How does one explain love? Stanley advised that it is the attraction that is important, not the how or why. We should be as unconcerned about the mystical attracting force of love as the magnet is unconcerned about ". . . how he / Doth th'obsequious iron move." His message is simple; accept love, do not question it.

The magnet and iron were used by William Schwenck Gilbert (1836-1911) to suggest an entirely different thought. His message is directed to one of man's basic dilemmas -- achieving one's potential. Bernard Gilpin said it most succinctly: "If it be right, do it boldly, -- if it be wrong leave it undone." We are also familiar with the exhortations of poets and writers that we set our sights high.

> Hitch your wagon to a starr.
>   - Ralph Waldo Emerson -

> Few things are impossible to diligence and skill.
>   - Samuel Johnson -

> Every noble work is at first impossible.
>   - Thomas Carlyle -

Gilbert departs from the rampant enthusiasm we have come to expect and accept. He points out that every man has limitations over which he has no control, and that it is foolish to continue in an endeavor that is beyond one's grasp. Unfortunately, Gilbert never suggests how one is to know the difference between what is within one's potential and what is not. As Shakespeare has so adequately expressed, "We know what we are, but know not what we may be."

> A magnet hung in a hardware shop,
> And all around was a loving crop
> Of scissors and needles, nail and knives,
> Offering love for all their lives;
> But for iron the magnet felt no whim,
> Though he charmed iron, it charmed not him;
> From needles and nails and knives he'd turn.
> For he'd set his love on a Silver Churn:
>
>> His most aesthetic,
>> Very magnetic
>> Fancy took this turn--
>> "If I can wheedle
>> A knife or a needle,
>> Why not a Silver Churn?"
>
> And Iron and Steel expressed surprise,
> The needles opened their well-drilled eyes,
> The penknives felt "shut up," no doubt,
> The scissors declared themselves "cut out,"
> The kettles boiled with rage, 'tis said,
> While every nail went off its head,
> And hither and thither began to roam,
> Till a hammer came up -- and drove them home.
>
>> While this magnetic,
>> Peripatetic
>> Lover he lived to learn,
>> By no endeavor

> Can a magnet ever
> Attract a Silver Churn!
> - "The Fable of the Magnet and the Churn" -
> - William Schwenck Gilbert -

Whereas we have seen how poets have used iron and magnets to construct analogies for the basis of their poems of love and faith, the twentieth century poet, May Sarton, used the most common of substances for one of hers -- water and salt.

> Consider the mysterious salt:
> In water it must disappear.
> It has no self. It knows no fault.
> Not even sight may apprehend it.
> No one may gather it or spend it.
> It is dissolved and everywhere.
>
> But out of water into air
> It must resolve into a presence,
> Precise, and tangible, and here.
> Faultlessly pure, faultlessly white,
> It crystallizes in our sight
> And has defined itself to essence.
>
> What element dissolves the soul
> So it may be both found and lost,
> In what suspended as a whole?
> What is the element so blest
> In which identity can rest
> As salt in the clear water cast?
>
> Love in its early transformation,
> And only love may so design it
> That the self flows in pure sensation,
> Is all dissolved and found at last

Without a future or a past,
And a whole life suspended in it.

The faultless crystal of detachment
Comes after, cannot be created
Without the first intense attachment.
Even the saints achieve this slowly;
For us, more human and less holy,
In time like air is essence stated.
                    - "In Time Like Air" -
                    - May Sarton -

She begins by developing the image of salt dissolving in water. The physical presence of the salt is lost; it no longer appears to exist. She then causes us to visualize its re-creation as the air slowly causes the water to evaporate leaving in its place a substance "Faultlessly pure, faultlessly white, / It crystallizes in our sight." The salt does exist; in the air it was slowly reborn. Sarton then asks "What element dissolves the soul / So it may be both found and lost?"

Love is the "element so blest" that it holds one's true self -- spirit and soul -- in a delicate state of "intense attachment." The salt crystallizes in air to "resolve into presence"; whereas the true self crystallizes in time to free the spirit and soul.

We have considered value in several of its forms -- fame and gold for example. We have focused on the visible and obvious aspects, yet true value exists more often than not in that which is more common and easily taken for granted. Joan Swift, a contemporary poet, adds another dimension to the concept of value. She tells us in her poem, "Oxygen," that true value resides in the most common element of our existence. Before reading it, we might wonder what more could be said about the gas other than it is colorless, odorless and tasteless? Perhaps the words she uses to express these properties leave deeper impressions.

Bearer of finches and clouds, pale atmosphere
Holds it, a rose in a bouquet of daises,
Although odorless, the one-fifth of each breath
That keeps flesh firm on the bone, the old blood warm,

Thought bright as young fish in the brain. Botanists
Say that plants exhale this element, push it
Out though olive skin. But who can say whether
It is this calm expiration or merely

Twilight and the first dew when, bending above
Vinca, hovering over viburnum leaves,
Underneath wide maples, we fill our nostrils
With a cool abundance of what gives us life?
- "Oxygen" -
- Joan Swift -

 Swift creates a forgotten appreciation for this element of life. We take note of how effortlessly she reminds us of this ocean in which we are immersed, of our physiological dependence on it, and of its photosynthetic regeneration. She questioningly suggests that it could be this substance alone that provides us with an exuberance and tranquility. But perhaps she also suggests that it is the totality of experience that is the "cool abundance of what gives us life" since life is more than mere physiological existence.
 But man does not live by serious thoughts alone. We must not leave our overview of chemistry and the poet without looking at one of the many "just for fun" poems. Does it have a message? Probably . . . . not.

When hydrogen played oxygen,
And the game had just begun,
Hydrogen racked up two fast points,
But oxygen had none.

Then oxygen scored a single goal,
And thus it did remain.
Hydrogen 2 and oxygen 1,
Called off because of rain.
> - "When Hydrogen Played Oxygen" -
> - anonymous -

CHAPTER V LIGHT

> I do not know what I may appear to the world; but to myself I seem to have been only a boy playing on the sea-shore... whilst the great ocean of truth lay all undiscovered before me.
>   -Sir Isaac Newton-
>   -from *Memoirs of Newton*-

"Light ceased to be an all-pervading mysterious substance of colourless purity, the very dwelling-place of God, and became a physical manifestation, having laws to be investigated with mirrors and lenses, and colours to be analysed by a prism."[1] So spoke Sir William Cecil Dampier. Not since Alexander the Great had a young man in his early twenties produced as significant an impact upon the western world as did Isaac Newton in the early eighteenth century. In the first chapter of the Book of Genesis we find the verse:

3. And God said, "Let there be light: and there was light.
4. And God saw the light and it was good:

In the "Epitaph Intended for Sir Isaac Newton, " we read:

> Nature and Nature's laws lay hid in night;
> God said, Let Newton be! and all was light.
> - Alexander Pope -

We find it difficult today to appreciate the impact that Newton's book, *Opticks*, had upon eighteenth century western Europe. For scores of centuries natural phenomena had been accepted as mystical workings of God and as beyond the understanding of man. Then in 1704, after spending a year and a half in a self imposed exile to avoid the ravages of the plague, Newton announced his discoveries of the nature and characteristics of light. Not only did he explain the origin and properties, but also described them mathematically and replicated them in the laboratory! For some people, his discoveries represented a new illumination in man's search for knowledge; for others, they represented a plunge into a darkness of despair brought about by a harsh, impersonal, mechanical explanation of what had once been mystical and heavenly.

William Powell Jones, in his book *The Rhetoric of Science*, writes of the poetic tributes paid to Newton during the latter part of the early 1800s.

> Newton the volume of skie unseals,
> And all th' amazing miracle reveals.
>   *   *   *   *
> Newton without rival alone,
> Prince of the new philosophy, his own.
>   *   *   *   *
> The works of nature, that in embryo lay,
> Drawn into life, and in a flood of day
> Newton's great genius to the world convey.[2]

Science would never again be the same following Newton's work, nor would philosophy, nor would the language, nor would art. The discoveries of the properties of light had a profound effect upon each of them. The theories of light removed one of the last bastions of mysticism from the sciences. The interventions of God were now recognized by philosophers as patterns of order rather then as aberrations. Colors became more a part of the written and spoken vocabularies. And in art, the painters used colors the way they were actually perceived by the artist rather that in the romanticized coloration of the accepted style. We must note, however, that artists were somewhat slower to make changes than were the poets and philosophers, The innovative styles of Seurat, Van Gogh, and Toulouse-Lactrec were not to peak until the latter part of the nineteenth century. Meanwhile, the conservative Puritan found a means to encourage moderation even among the painters and their use of color. James Cawthorn (1719-1771) used the relationship between color and light to moralize in his poem, "Regulation of the Passions."

> Passions, like colours, have their strength and ease,
> Those too insipid, and too gaudy these:
> Some on the heart, like Spagnoletti's throw
> Fictitious horrours, and a weight of woe;
> Some, like Albano's, catch from ev'ry ray
> Too strong a sunshine, and too rich a day;
> Others, with Carlo's Magdalens, require
> A quicker spirit, and touch of fire...
> Wouldst thou then reach what Rembrandt's genius knew,
> And live the model that his pencil drew,
> From all thy life with all his warmth divine,
> Great as his plan, and faultless as his line;
> Let all thy passions, like his colours, play,
> Strong without harshness, without glaring gay.
>   -from "The Regulation of Passions"-
>   -James Cawthorn-

Newton's new science proved what had been intuitive. The universe, and all of its components, was a design of ultimate order. Philosophers and religious leaders took comfort in the reality. Yet a nagging disquiet continued. In some ways the mathematical, mechanical modeling had reduced the mystery of the universe and of creation, and had diluted God's magnificence. Sir William Cecil Dampier summarized this unsettling feeling.

> It is evident that in the age of Newton - the age of the first great synthesis of scientific knowledge - the revolution in intellectual outlook of mankind involved a revolution in the statement of dogmatic religious belief...
>
> Light ceased to be an all-pervading mysterious substance of colourless purity, the very dwelling-place of God, and became a physical manifestation, having laws to be investigated with mirrors and lenses, and colours to be analyzed by prisms.[3]

As a result, certain of the poets devoted their efforts to attempting to demonstrate that the increasing knowledge gained from science was in fact more evidence of the glory of God.

In William Cowper's (1731-1800) long six part poem, "The Task," he specifically uses the new knowledge of light to praise God. First he chastises man for taking God's creation for granted, "...The landscape has his praise, / But not its author...." Man appreciates the beauty and wonder of nature, he is saying, but forgets it was created by God.

> Acquaint thyself with God if thou would'st taste
> His works. Admitted once to his embrace,
> Thou shalt perceive that thou wast blind before:
> Thine eye shall be instructed, and thine heart
> Made pure, shall relish, with divine delight
> 'Til then unfelt, what hands divine have wrought.
> Brutes graze the mountain-top with faces prone
> And eyes intent upon the scanty herb
> It yields them, or recumbent on its brow,

Ruminate heedless of the scene outspread
Beneath, beyond, and stretching far away
From inland regions to the distant main.
Man views it and admires, but rests content
With what he views. The landscape has his praise,
But not its author. Unconcern'd who form'd
The paradise he sees, he finds it such,
And such well-pleased to find it, asks no more.
Not so the mind that has been touch'd from heav'n,
And in the school of sacred wisdom thought the world,
To read his wonders, in whose thought the world,
Fair as it is, existed ere it was.
Not for its own sake merely, but for his
Much more who fashioned it, he gives it praise;
Praise that from earth resulting as it ought,
To earth's acknowledg'd sov'reign, finds at once
Its only just proprietor in Him.
The soul that sees him, or receives sublimed
New faculties, or learns at least t' employ
More worthily the pow'rs she own'd before;
Discerns in all things, what with stupid gaze
Of ignorance, till then she overlook'd,
A ray of heav'nly light gilding all forms
Terrestial, in the vast and the minute
The unambiguous footsteps of the God
Who gives the lustre to an insect's wing,
And wheels his throne upon the rolling worlds.
Much conversant with heav'n, she often holds
With those fair ministers of light to man,
That fill the skies nightly with silent pomp,
Sweet conference.
     - from "The Task:
       Winter Morning Walk" -
   - William Cowper -

Cowper reiterates his thesis more positively in Book IV. He

departs from a direct reference to light, but dwells upon the idea that the orderly universe was created by God and that once it was set into motion it continued according to the design "And need not his immediate hand, who first / Prescribed their course, to regulate it now." He continues to emphasize that the knowledge acquired through science is only further recognition of God's laws: "Nature is but a name for an effect / Whose cause is God . . ."

> Some say that in the origin of things
> When all creation started into birth,
> The infant elements received a law
> From which they swerve not since. That under force
> Of that controuling ordinance they move,
> And need not his immediate hand, who first
> Prescribed their course, to regulate it now.
> Thus dream they, and contrive to save a God
> The incumbrance of his own concerns, and spare
> The great Artificer of all that moves
> The stress of a continual act, the pain
> Of unremitted vigilance and care,
> As too laborious and serve a task.
> So man the moth is not afraid, it seems,
> To span Omnipotence, and measure might
> That knows no measure, by the scanty rule
> And standard of his own, that is today,
> And is not, ere to-morrow's sun go down.
> But how should matter occupy a charge
> Dull as it is, and satisfy a law
> So vast in its demands, unless impell'd
> To ceaseless service by a ceaseless force,
> And under pressure of some conscious cause?
> The Lord of all, himself through all diffused,
> Sustains and it is the life of all that lives.
> Nature is but a name for an effect
> Whose cause is God. He feeds the secret fire
> By which the mighty process is maintain'd,

Who sleeps not, is not weary; in whose sight
Slow-circling ages are as transient days;
Whose work is without labor, whose designs
No flaw deforms, no difficulty thwarts,
And whose beneficence no charge exhausts.
> - from "The Task:
> Winter Walk at Noon" -
> - William Cowper -

The romantic poet would take the opportunity at times to incorporate into the body of his work a protest against the new science. An example of this approach is found in the long poem, "Lamia," written in 1819 by John Keats. From Part II we find the following:

What wreath for Lamia? What for Lycius?
What for the sage, old Apollonius?
Upon her aching forehead be there hung
The leaves of willow and of adder's tongue;
And for the youth, quick, let us strip for him
The thyrsus, that his watching eyes may swim
Into forgetfulness; and, for the sage,
Let spear-grass and the spiteful thistle wage
War on his temples. Do not all charms fly
At the mere touch of cold philosophy?
There was an awful rainbow once in heaven:
We know her woof, her texture; she is given
In the dull catalogue of common things.
Philosophy will clip an Angel's wings,
Conquer all mysteries by rule and line,
Empty the haunted air, and gnomed mine--
Unweave a rainbow, as it erewhile made
The tender-person'd Lamia melt into a shade.
> - from "Lamia II" -
> - John Keats -

At line 229 we find the rhetorical question: "Do not all charms fly/ At the mere touch of cold philosophy?" We must keep in mind that his use of philosophy refers to what we would call science. He goes on to say that science, particularly because of Newton, has "unweaved" the rainbow. Newton has "destroyed all the poetry of the rainbow by reducing it to the prismatic colours." He further criticizes Newton for degrading the magic of the rainbow to the point that it becomes only an entry "in the dull catalog of common things." Professor Earl R. Wasserman in his book, *The Finer Tone*, writes that in "Lamia," Keats "was facing squarely the unpleasant fact that mortal man cannot avoid the conditions that allow the conceptual mind to intrude and:

> 'Conquer all mystery by rule and line,
> Empty the haunted air, and gnomed mine -
> Unweave a rainbow, as it erstwhile made
> The tender-person'd Lamia melt into shade'."[4]

More subtle uses of light as an ingredient in poetry should be mentioned. Two uses are exhibited by Emily Dickinson, the quiet, delicate poet of nineteenth century New England. Each one creates a mood and expresses to a degree a kind of reverence. Each one defines a limit to scientific knowledge, and in essence says that more exists to light than can be measured, reflected or refracted by science.

> There's a certain slant of light,
> On winter afternoons,
> That oppresses, like the weight
> Of cathedral tunes.
>
> Heavenly hurt it gives us;
> We can find no scar.
> But internal difference
> Where the meanings are.

None may teach it anything
"Tis the seal, despair,--
An imperial affliction
Sent us of the air.

When it comes, the landscape listens,
Shadows hold their breath;
When it goes, 'tis like the distance
On the look of death.
             -"There's a Certain Light"#258-
             -Emily Dickinson-

As a footnote to this poem it should be mentioned that psychologists and physiologists have strong evidence that certain of people's moods are physiologically associated with particular kinds of light. The study of depression, as an example, has provided indications that some people are directly affected by the quality and quantity of light to which they are exposed. The Artic explorer Frederick Cook[5] noted in his expedition's journal in May, 1898:

> The winter and darkness have slowly but steadily settled over us....It is not difficult to read on the faces of my companions their thoughts and their moody dispositions....The curtain of blackness which has fallen over the outer world of icy desolation has also descended upon the inner world of our souls....all efforts to infuse bright hopes fail.

We recognize today that Cook's men were suffering from the psychiatric disease known as seasonal affective disorder. It now appears that the disorder is probably caused by the biochemical system involving the hormone melatonin, which affects mood and subjective energy levels. As for light, scientists are finding that indeed "Heavenly hurt it gives us."

The second poem, "A Light Exists in Spring," can be

interpreted as Dickinson's reminder that science has limits that the human senses have not.

> A Light exists in Spring
> Not present on the Year
> At any other period ---
> When March is scarcely here
>
> A Color stands abroad
> On Solitary Fields
> That Science cannot overtake
> But Human Nature feels.
>
> It waits upon the Lawn,
> It shows the furthest Tree
> Upon the furthest Slope you know
> It almost speaks to you.
>
> Then as Horizons step
> Or Noons report away
> Without the Formula of sound
> It passes and we stay ---
>
> A quality of loss
> Affecting our Content
> As Trade had suddenly encroached
> Upon a Sacrament.
>
>   - "A Light Exists in Spring" #812 -
>   - Emily Dickinson -

In the second stanza Dickinson prompts us to recall the special light we have experienced that casts its spell over the early spring day. We have seen it and felt it but ". . . Science cannot overtake." Her thought is that this light, like many other phenomena, is to be experienced but not measured. It can be encountered, but not proven.

Robert Frost's poem, "For Once, Then, Something," deserves special attention. The critic Lawrence Thompson wrote: "At first glance, the central image of an action represents only the familiar rural pastime of trying to look down through the water, in a well, to see the bottom, or to see how deep the well is. Yet the metaphorical undertones and metaphysical overtones are cunningly interwoven."[6]

Frost, as a careful observer of natural phenomena, describes correctly the nature of the images within the context of the scientific principles of light. Frost refers to the speaker's physical positioning "Always wrong to the light . . . ." The brightness of the unrestricted sky light is reflected directly to him from the surface of the water. The image he sees is the sky and "cloud puffs." The light bouncing from his face naturally hits the surface almost vertically and is therefore reflected directly back with the result that ". . . the water gives me back in a shining surface picture." Thus we have two images produced from the water's surface. His own face and the bright, extraneous environment.

> Others taunt me with having knelt at well-curbs
> Always wrong to the light, so never seeing
> Deeper down in the well than where the water
> Gives me back in a shining surface picture
> Me myself in the summer heaven godlike
> Looking out of a wreath of fern and cloud puffs.
> Once, when trying with chin against a well-crub
> I discerned, as I thought, beyond the picture,
> Through the picture, a something white, uncertain,
> Something more of the depths--and then I lost it.
> Water came to rebuke the too clear water.
> One drop fell from a fern, and lo, a ripple
> Shook whatever it was lay there at bottom,
> Blurred it, blotted it out. What was that whiteness?
> Truth? A pebble of quartz? For once, then, something.
>   - "For Once, Then, Something" -
>   - Robert Frost -

Critic Jeffrey Hart interprets the image of the kneeling viewer ". . . godlike / Looking out of a wreath of ferns and cloud puffs." as "an anthromorphic god projected by man's vanity into the heavens." Or the view projected "may be humanism of the sort that contains in its conspectus nothing higher than man."[7] Considering Hart's interpretation, we might revise the biblical statement that "God created man in *his* own image", to that of, "Man created god in *his* own image."

However, Frost began by describing himself as always being wrong to the light and being taunted for never seeing the bottom. Scientifically he is correct. His positioning allowed the brilliant sky light to overpower the more dimly reflected light from the bottom, thus inhibiting its being seen. He sees the bright, gaudy peripheral and himself; they capture his attention and distract his sight from the deeper importance. We are left to infer that the only way to see, in depth, to the bottom, is to block out with ourselves the brighter distractions, and to focus our attention on the depths rather than on the superficial self.

Frost leaves us with one other unanswered question. Who are the "others (who) taunt me"? Are there "others," or is it himself -- reproaching his own vanity and imperfection?

Carl Sandburg has been described as the first poet of modern times actually to use the language of the people as his almost total means of expression.[8] In 1936, he wrote *The People, Yes,* a book-length narrative poem which has been characterized as a combination catalog and commentary. Professor Richard Crowder, a scholar of American literature, describes it as a "catch-all of folklore and rhapsodic flights." It consists of proverbs, anecdotes, folk-wisdom, dialogs -- all against a background of observations and prophecies in Sandburg's own idiom, sometimes mystical, sometimes slangy."[9]

In one respect the structure resembles somewhat the composition of William Blake's "Auguries of Innocence" which we examined in our chapter concerning the atom. Blake included a rather lengthy sucession of proverbs known to the people of his time. Sandburg, in *The People, Yes*, provides much of the same sort of insight into the folklore of his period. Not only do we find a comparison of this commonality interesting, but also a comparison of the language and styles is fascinating. Only the last section of the poem is quoted here, but it inclds two significant uses of the scientific principles of light.

>The people will live on.
>The learning and blundering people will live on.
>They will be tricked and sold and again sold
>And go back to the nourishing earth for rootholds,
>The people so peculiar in renewal and comeback,
>You can't laugh off their capacity to take it.
>The mammoth rests between his cyclonic dramas.
>
>The people so often sleepy, weary, enigmatic,
>is a vast huddle with many units saying:
>>"I earn my living.
>>I make enough to get by
>>and it takes all my time.
>>If I had more time
>>I could do more for myself
>>and maybe for others.
>>I could read and study
>>and talk things over
>>and find out about things.
>>It takes time.
>>I wish I had the time."
>
>The people is a tragic and comic two-face:
>hero and hoodlum: phantom and gorilla twist-
>ing to moan with a gargoyle mouth: "They
>buy me and sell me . . . it's a game . . .
>sometime I'll break loose . . ."

       Once having marched
Over the margins of animal necessity,
Over the grim line of sheer subsistence
       The man came
To the deeper rituals of his bones,
To the lights lighter than any bones,
To the time for thinking things over,
To the dance, the song, the story,
Or the hours given over to dreaming,
       Once having so marched.

Between the finite limitations of the five senses
and the endless yearnings of man for the beyond
the people hold to the humdrum bidding of work and food
while reaching out when it comes their way
for lights beyond the prisms of the five senses,
for keepsakes lasting beyond any hunger or death.
     This reaching is alive.
The panderers and liars have violated and smutted it.
     Yet this reaching is alive yet
     for lights and keepsakes.

          The people know the salt of the sea
          and the strength of the winds
          lashing the corners of the earth.
          The people take the earth
          as a tomb of rest and a cradle of hope.
          Who else speaks for the Family of Man?
          They are in tune and step
          with constellations of universal law.

      The people is a polychrome,
      a spectrum and a prism
      held in a moving monolith,
      a console organ of changing themes,
      a clavilux of color poems

wherein the sea offers fog
and the fog moves off in rain
and the labrador sunset shortens
to a nocturne of clear stars
serene over the shot spray
of northern lights.

The steel mill sky is alive.
The fire breaks white and zigzag
shot on a gun-metal gloaming.
Man is a long time coming
Man will yet win.
Brother may yet line up with brother:

This old anvil laughs at many broken hammers.
    There are men who can't be bought.
    The fireborn are at home in fire.
    The stars make no noise.
    You can't hinder the wind from blowing.
    Time is a great teacher.
    Who can live without hope?

In the darkness with a great bundle of grief
    the people march.
In the night, and overhead a shovel of stars for
    keeps, the people march:
                    "Where to? what next?"
          - from *The People, Yes* #107 -
          - Carl Sandburg -

    We should give special attention to the six lines beginning with "Between the finite limitations of the five senses." In the fifth line of that part we find, "for lights beyond the prisms of the five senses." In a later version of the poem, we find that one word has been changed -- prism becomes prison. Notice that both words have the same number of syllables and have the same basic sound, but what a difference in meaning!

"Between the limitations of the five senses" (which we can call the domain of the known) and "the endless yearnings of man for the beyond" (the desire for understanding the unknown), man plods forward opportunely reaching out and closing the gap between the known and the unknown. The prisms of his senses serve as a metaphor for man's intellectual processes. His intellect permits him to expand the effects of his experiences -- to open them, to spread them, to analyze them -- as a prism separates and displays the solar spectrum allowing for its analyses. The prisms of the senses permit man to create, to invent, and to reach out for the beyond. Man is not tied to the literal stimulation of his senses only. He deciphers, translates, and interprets the stimuli for his own purpose and for a grander purpose. He reaches out, "for lights beyond the prisms."

Another reading of the poem encourages us to consider his use of the senses as analogous to reality, and his "reaching out for the lights," as the struggle for ideals. Again, the prisms of our senses allow us to look beyond the reality of the moment, and to reach for the ideals which are, "the keepsakes lasting beyond any hunger or death."

When we consider the second version of these six lines, the version that changes prisms to prisons, we see a pronounced change in the meaning. "The prisons of the five senses," when used metaphorically, tells us that man's senses restrain his intellect; they imprison him to narrowly defined limits, to old patterns, to old stereotypes. They inhibit creativity and invention. Man is, however, persistent. In spite of his "finite limitations," he continues to reach out for "the lights beyond the prisons." In spite of his humdrum plodding, ". . . this reaching is alive yet."

So we see that in one version, Sandburg says that man's senses are like prisms -- enabling him to reach beyond himself. In the other version, Sandburg says that man's senses are like prisons -- chaining him to harsh reality, nevertheless he still reaches beyond. In either version Sandburg's dominant theme persists -- "man will yet win."

Sandburg concludes *The People, Yes* by describing people as a polychrome and a moving monolith -- a seeming contradiction. His monolith -- the massive undifferentiated whole, one harmonious pattern throughout -- is actually comprised of an infinite variety of discrete components. This is the society of man. A society not unlike light itself. Newton's white light is perceived as an undifferentiated whole, but in fact it is composed of an overwhelming variety of discrete and different photons. It is polychromatic.

Sandburg says that the people is, "a spectrum and a prism," and thus confronts us with what seems to be a contradiction. A prism is a producer, and a spectrum is a product. Just as a prism can bend the path of polychromatic light, the people can redirect the energy of the moving monolith (society) -- "a console organ of changing themes." Just as a spectrum displays the colors within the white light, the people display the constituents of its monolithic (societal) being -- its hopes, its griefs, its aspirations, its strengths, its wisdom, and even its "blundering."

Sandburg's message reiterates his conviction: "The people will live on." The question is not, will they? The only question is, "Where to? what next?"

In the darkness . . . the people march
In the night . . .the people march.

CHAPTER VI   SOUND

> The great tragedy of Science -- the slaying of a
> beautiful hypothesis by an ugly fact.
> - from *Biogenesis and Abiogenesis* -
> - T. H. Huxley -

Poetry is full of sounds. The words themselves are sounds; the rhythms are sounds; the messages are sounds. Poetry *is* sound.

>   Jubilant the music through the fields
>   a-ringing,--
>   Carol, warble, whistle, pipe, -- endless ways
>   of singing,
>   Oriole, bobolink, melody of thrushes,
>   Rustling trees, hum of bees, sudden little
>   hushes
>   Broken suddenly again --
>   Carol, whistle, rustle, humming,
>   In reiterate refrain,
>   Thither, hither, going, coming,
>   While the streamlets' softer voices mingle

murmurously together;
Gurgle, whisper, lapses, splashes,--praise
    of love and summer weather.
Hark! A music finer on the air is blowing--
Throbs of infinite content, sounds of things
    a-growing,
Secret sounds, flit of bird under leafy cover,
Odors shy floating by, clouds blown swiftly over,
Kisses of the crimson roses,
Crosses of the lily-lances,
Stirrings when a bud uncloses,
Tripping sun and shadow dances,
Murmur of aerial tides, stealthy zephyrs gliding,
Far yet clear, strange yet near, sweet
    with a profounder sweetness,
Mystical, rhythmical, weaving all into
    completeness;
For its wide, harmonious measures
Not one earthly note let fall;
Sorrows, raptures, pains and pleasures,
All in it, and it in all,
Of earth's music the ennobler, of its discord
    the refiner,
Pipe of Pan was once its naming, now it
    hath a name diviner.
And a thousand nameless things sweeter
    for their hiding.
Ah! a music more than these floweth on
    forever,
In and out, yet beyond our tracing or
    endeavor,
        -"World Music"-
        -Frances Louisa Bushnell-

This poem, "World Music," draws us to an earlier idea -- what the writer Arthur Koestler (1905-1983) called the "cosmic music box."

The cosmic music box

A prevailing conception throughout history until the early seventeenth century can best be described as the idea of the music of the spheres. Pythagoras and his followers believed the universe was ruled by harmony which translated into numbers and music. All of the different heavenly revolutions produced different tones. Thus, each of the planets and spheres of the fixed stars produced its own unique musical sound. The major problem with the idea was that no one could hear them. Undaunted in their belief, the Pythagoreans explained that man was unable to hear the sounds because he had heard them since birth and had, therefore, become accustomed to the music -- but it did exist because Pythagoras himself was able to hear the music. (Aristotle took exception to the "music of the spheres" because he said that the celestial bodies were so large that the noise would be deafening. And besides, the bodies did not move!)

Nevertheless, even though the premise was weak, most philosophers, theologians, and poets were so enamored with the idea of heavenly music and the idea that harmony ruled the universe, that it held their minds captive. For as John Donne wrote in his "Sermons," "God made the whole world in such a uniformity, such a concinnity of parts, as that it was an Instrument, perfectly in tune."

It was Johannes Kepler who, in his 1618 book entitled *Harmonice Mundi (Harmony of the World),* attempted to display the secret of the universe in an all embracing synthesis of geometry, music, astrology, astronomy, and epistemology. Kepler utilized the data collected by Tycho de Brahe (1546-1601) which described the paths, times, and distances of the planets. He computed an exhaustive variety of arithmetical combinations of the data searching for the numerical relationships he intuitively knew to be present. He had found geometric proportions everywhere in nature, and he considered them to be the pure harmonies which guided God in the work of Creation. The sensory harmony, our recognition of sound as music, he considered to be only an echo of it.

Finally in 1618 he discovered that variations in the angular velocities of the planetary motion, as viewed from the sun, regardless of the distance, produced the whole number ratios he sought. He was able to convery these ratios into musical scales particular to each of the planets. The celestial harmonies -- the "musick of the spheares" -- had at last been found. But Kepler wrote quite clearly that:

> For there are two things which disclose harmonies in natural things: either light or sound ... there are no such sounds in the heavens, nor is movement so turbulent that any noise is made by the rubbing against the ether. Light remains...Harmonies are visual not sound.[9]
> -from *Harmonice Mundi* Book IV-
> -Johannes Kepler-

Kepler demonstrated that harmony was only a mathematical conception which one could see but could not hear. One would think, therefore, that his work would have ended the idea of the heavenly music. But its romance was so attractive that even Kepler had difficulty in divorcing himself from it.

> In the Celestial Harmonies, which planet sings soprano, which alto, which tenor, and which bass? Although these words are applied to human voices while voices and sounds do not exist in the heavens... I do not know why but nevertheless this wonderful congruence with human song had such a strong effect upon me that I am

compelled to pursue this part of the comparison, also, even without any solid natural cause.[10]

>   -from *Harmonice Mundi* Book V-
>   -Johannes Kepler-

And true to his word, Kepler wrote that by his calculations if several planets are simultaneously at the extreme points of their respective orbits, the result is a musical composition in which Saturn and Jupiter represent the bass, Mars the tenor, Earth and Venus the contralto, and Mercury the soprano.

With this brief history as the background, it is easy for us to understand the Pythagorean influence -- the musick of the sphears -- up to the seventeeth century. Typical of the uses of the concept by the poet is the reference made by the poet Sir Phillip Sidney (1551-1586) as the character Klaius describes one of the female characters:

> For she, with whom compar'd, the Alps are vallies,
> She, whose lest word brings from the spheares their musique,
> At whose approach the Sunne rase in the evening,
> Who, where she went, bare in her forehead morning,
> Is gone, is gone from these our spoyled forrests,
> Turning to desarts our best pastur'de mountaines.
>   -from "IVth Ecloques"-
>   -Sir Philip Sidney-

Yet even he expressed cautious doubt, at times, about the reality of this music; "If the senceless spheares doo yet hold a musique."[11]

In Shakespeare's *Merchant of Venice* we find in Act V, line 60: "There's not the smallest orb which thou beholdest / But in his motion like an angel sings." And in *Twelfth Night*, Olivia expresses her wish for the conversation with Viola to continue by saying:

> O, by your leave, I pray you,
> I bade you never speak again of him:
> But, would you undertake another suit,
> I had rather hear you to solicit that
> Than music from the spheres.
>
> -from *Twelfth Night* Act III, 117-121-
> -William Shakespeare-

Still, even after Kepler' *Harmonice Mundi*, we find that the idea of celestial music persisted longer than one might think. One reason was because of the slow rate of the diffusion of Kepler's work, principally because only a limited number of people had access to it and an understanding of its implications. Another reason was that the idea of heavenly music was simply too enthralling to be discarded. It described, better that any other concept, the beauty, glory, and perfection of God's Creation -- even though science had shown it to be untrue.

John Donne, whose life straddled both sides of *Harmonice Mundi*, accepted the revelations of Kepler. We find these lines in his poem "Upon the Translation of the Psalms by Sir Philip Sidney;" "The Spheares have Musick, but they have no tongue, / Their harmony is rather danc'd than sung." Donne, in essence, repeated Kepler's rather direct statement that "Harmonies are visual not sound."

George Herbert (1593-1633), another of the metaphysical poets at this time of transition, also adopted Kepler's soundless harmony. Yet, he went one step further. In his poem "Artillerie," in which he verbally jousts with God over the missles fired by each at the other, he inserts a mild joke in response to a fallen star which has spoken to him. This modest deceit in the ninth line is one of the first to be directed at the holy musick.

> As I one ev'ning sat before my cell,
> Me thoughts a starre did shoot into my lap.
> I rose, and shook my clothes, as knowing well,

That from small fires comes oft no small mishap.
> When suddenly I heard one say,
> Do as thou usest, disobey,
> Expell good motions from thy breast,
Which have the face of fire, but end in rest.

I, who had heard of musick in the spheres,
But not of speech in starres, began to muse:
The starres and all things are; If I refuse,
> Dread Lord, said I, so oft my good;
> Then I refuse not ev'n with bloud
> To wash away my stubborn thought:
For I will do or suffer what I ought.

But I have also starres and shooters too,
Born where thy servants both artilleries use.
My tears and prayers night and day do wooe,
And work up to hee; yet thou dost refuse.
> Not but I am (I must say still)
> Much more oblig'd to do thy will,
> Then thou to grant mine: but because
Thy promise now hath ev'n set thee thy laws.

Then we are shooters both, and thou dost deigne
To enter combate with us, and contest
With thine own clay. But I would parley fain:
Shunne not my arrows, and behold my breast.
> Yet if thou shunnest, I am thine:
> I must be so, if I am mine.
> There is no articling with thee:
I am but finite, yet thine infinitely.
> - "Artillerie" -
> - George Herbert -

Joseph Addison is another of the poets who represent those who accepted the new explanation. In his "Hymn," which is also known as "Spacious Firmament on High" and which is previously quoted in this book, the last stanza is particularly meaningful. He basically says -- So what? The absence of physical sound does not diminish the majesty of the universe.

> The spacious firmament on high,
> With all the blue ethereal sky,
> And spangled heavens, a shining frame,
> Their great Original proclaim.
> The unwearied Sun from day to day
> Does his Creator's power display;
> And publishes to every land
> The work of an Almighty hand.
>
> Soon as the evening shades prevail,
> The Moon takes up the wondrous tale;
> And nightly to the listening Earth
> Repeats the story of her birth:
> Whilst all the stars that round her burn,
> And all the planets in their turn,
> Confirm the tidings as they roll,
> And spread the truth from pole to pole.
>
> What though in solemn silence all
> Move round the dark terrestrial ball;
> What though nor real voice nor sound
> Amidst their radiant orbs be found?
> In Reason's ear they all rejoice,
> And utter forth a glorious voice;
> Forever singing as they shine,
> "The Hand that made us is divine."
>   - from "Spacious Firmament on High" -
>   - Joseph Addison -

Some of the seventeenth century poets seemed uncertain as whether to retain or to discard the ancient idea. John Milton

(1608-1674) provides an example of the vacillation in his poem "Arcades." At one point he listens to the "Celestial siren's harmony," but later states that no human ear can hear it.

> For know by lot from Jove I am the power
> Of this fair wood, and live in oaken bower,
> To nurse the saplings tall, and curl the grove
> With ringlets quaint and wanton windings wove;
> And all my plants I save from nightly ill
> Of noisome winds and blasting vapors chill;
> And from the boughs brush off the evil dew,
> And heal the harms of thwarting thunder blue,
> Or what the cross dire-looking planet smites,
> Or hurtful worm with cankered venom bites.
> When evening gray doth rise, I fetch my round
> Over the mount and all this hallowed ground,
> And early ere the odorous breath of morn
> Awakes the slumbering leaves, or tasseled horn
> Shakes the high thicket, haste I all about,
> Number my ranks, and visit every sprout
> With puissant words and murmurs made to less.
> But else in deep of night, when drowsiness
> Hath locked up mortal sense, then listen
> To the celestial sirens' harmony,
> That sit upon the nine infolded spheres
> And sing to those that hold the vital shears
> And turn the adamantine spindle round,
> On which the fate of gods and men is wound.
> Such sweet compulsion doth in music lie,
> To lull the daughters of Necessity,
> And keep unsteady Nature to her law,
> And the low world in measured motion draw
> After the heavenly tune, which none can hear
> Of human mold with gross unpurged ear;

And yet such music worthiest were to blaze
The peerless height of her immortal praise
Whose luster leads us, and for her most fit,
If my inferior hand or voice could hit
Inimitable sounds;
   -from "Arcades"-
   -John Milton-

Vestiges of the belief in the celestial music and heavenly harmony continued into the nineteenth century. Shelley wrote in "Scenes from the Faust of Goethe," that "The sun makes music as of old / Amid the rival spheres of Heaven." And in his classic work, "Prometheus Unbound," we find the dialog:

   CHORUS OF SPIRITS
  Our spoil is won,
  Our task is done,
We are free to dive, or soar, or run;
  Beyond and around,
  Or within the bound
Which clips the world with darkness round.

  We'll pass the eyes
  Of the starry skies
Into the hoar deep to colonize;
  Death, Chaos and Night,
  From the sound of our flight,
Shall flee, like mist from a tempest's might.

  And Earth, Air and Light,
  And the Spirit of Might,
Which drives round the stars in their fiery flight;
  And Love, Thought and Breath,
  The powers that quell Death,
Whereever we soar shall assemble beneath.

    And our singing shall build
    In the void's loose field
A world for the Spirit of Wisdom to wield;
    We will take our plan
    From the new world of man,
And our work shall be called the Promethean.

CHORUS OF HOURS
Break the dance, and scatter the song;
  Let some depart, and some remain;

SEMICHORUS I
We, beyond heaven, are driven along;

SEMICHORUS II
Us the enchantments of earth retain;

SEMICHORUS I
Ceaseless, and rapid, and fierce and free,
With the Spirits which build a new earth
    and sea,
And a heaven where yet heaven could never
    be;

SEMICHORUS II
Solemn, and slow, and serene, and bright,
Leading the Day, and outspeeding the
    Night,
With the powers of a world of perfect
    light;

IONE
    Yet fell you no delight
From the past sweetness?

PANTHEA

                    As the bare green hill,
When some soft cloud vanishes into rain,

Laughs with a thousand drops of sunny
        water
To the unpavilioned sky!

                IONE
                    Even whilst we speak
New notes arise. What is that awful sound?

                PANTHEA
'Tis the deep music of the rolling world,
Kindling within the strings of the waved air
Aeolian modulations.

                IONE
                Listen too,
How every pause is filled with under-notes,
Clear, silver, icy, keen awakening tones.
Which pierce the sense, and live within the
        soul,
As the sharp stars pierce winter's crystal
        air
And gaze upon themselves within the sea.
            - from "Prometheus Unbound" IV -
            - Percy Bysshe Shelley -

Today, as a particular church hymn is sung, one must wonder how many of the singers are aware of the history, the philosophy, the theology, the science, and the poetry that are contained in these four lines.

    This is my Father's world,
    And to my listening ear,

All nature sings and around me rings,
The music of the spheres.
- from "This Is My Father's World" -
- Matbie D. Babcock (1858-1901) -

The physics of sound

If we take our attention away from the abstractions of sound for a moment, we shall find that sometimes the poet goes beyond the pure lyrical considerations and uses the physical properties of sound to make his point. Robert Bly has done just that in his poem, "Watching Television."

The expression, "Sounds are heard too high for ears," refers to the electromagnetic waves radiating through the air bringing to the antenna and then to the television the audio and visual impulses. These electromagnetic waves are the sound and sight from the source, but the frequency is too high for our senses to respond. Other electronic devices and circuitry are necessary. We must have a converter, an amplifier, and some kind of translator, to adapt the phenomena to our senses' level of reception.

Bly describes the physiological process of the sound reception by us as, "From the body cells there is an answering bay." The auditory receptor cells receive the stimulus of the sound waves radiating from the television speaker system. Each neuron transmits the energy across its synapse to the next cell, and so on, until, "Soon the inner streets fill with a chorus of barks." A chain reaction of nerve impulses transmit the stimulus to the brain. Until this occurs -- until the brain receives the stimulus -- we have no true awareness or understanding of the enternal phenomena that has occurred. Until the conversion occurs, our bodies are bathed in the event, but we remain unaware and indifferent.

Bly follows his introductory stanza with a series of jumbled images from the television. Each is recognizable as some part of an event, but the pieces lack coherence. We absorb, but we do not process the stimulus. The complexity of the collage is too bothersome, "too high for ears," (for understanding), without a conversion and translation. Are we not willing to process intellectually the events that bombard us, or are we incapable so that with a shrug of indifference we withdraw and "The spirit breaks"? Is Bly telling us to listen; it's there; do something!

> Sounds are heard too high for ears,
> From the body cells there is an answering bay;
> Soon the inner streets fill with a chorus of barks.
>
> We see the landing craft coming in,
> The black car sliding to a stop,
> The Puritan killer loosening his guns.
>
> Wild dogs tear off noses and eyes
> And run off with them down the street--
> The body tears off its own arms and throws them into the air.
>
> The detective draws fifty-five people into his
>   revolver,
> Who sleep restlessly as in an air raid in London;
> Their backs become curved in the sloping dark.
>
> The filaments of the soul slowly separate:
> The spirit breaks, a puff of dust floats up,
> Like a house in Nebraska that suddenly explodes.
>   - "Watching Television" -
>   - Robert Bly -

Paul Simon has a similar theme in his poem, "The Sound of Silence."

> Hello darkness my old friend,
> I've come to talk to you again,
> Because a vision softly creeping,
> Left its seeds while I was sleeping,
> And the vision that was planted in my brain

Still remains within the sound of silence.

In restless dreams I walked alone
Narrow streets of cobble-stone,
'Neath the halo of a street lamp,
I turned my collar to the cold and damp
When my eyes were stabbed by the flash of a neon light
That split the night and touched the sound of silence.

And in the naked light I saw
Ten thousand people maybe more.
People talking without speaking,
People hearing without listening,
People writing songs that voices never share
And no one dare disturb the sound of silence.

"Fools!" said I, "You do not know
Silence like a cancer grows.
Hear my words that I might teach you,
Take my arms that I might reach you."
But my words like silent raindrops fell,
And echoed in the wells of silence.
           - "The Sound of Silence" -
           - Paul Simon -

Although Simon does not use the science of sound specifically, his call is the same as Bly's. Man is surrounded by events -- sights and sounds -- to which, whether by ignorance or indifference, he is oblivious. The physical phenomena of sight and sound exists, but he is unresponsive; he neither sees nor hears.

    People talking without speaking,
    People hearing without listening,

But the intelligibility of the physical phenomena requires processing by man. Without the processing all is " . . . echoed in the wells of silence."

All the sound has the same cause; it is a disturbance of the molecules of the medium. Energy at the origination of sound is transmitted through the air, or water, or wall, from one molecule to the next as a sort of rebounding-domino effect. Whether the sound is a symphony, a car crash, a sneeze, or a whisper, all are transmitted alike. Thus the question of whether it is music or noise, is in the ears of the beholder.

Because noise is so associated with civilization, the poets who are not so certain that man is civilizing himself often use sound/noise as the vehicle for their protests. The poem by the contemporary poet Rod McKuen is an example of this kind of protest and of the relativity of sound.

> If I had a pistol to hold in my hand
> I'd hunt down and silence the Good Humour man,
> I'd pour sticky ice cream all over his wound
> and stop him forever from playing that tune.
>
> For silence is golden on a soft summer day.
> It's a pity to let strangers take it away.
>
> If ever I get me a license to kill
> I'll war on the jukebox and jackhammer till
> the wind and the rain rust up all their parts
> and the worms and the woodchucks dissect their hearts.
>
> For silence is golden and hard to be found,
> and killed far too often by the jackhammer's sound.
> - "Silence Is Golden" -
> - Rod McKuen -

If McKuen is protesting the uncivilized behavior of his fellow man and his insensitive noisemaking, we can only wonder about the level of civilization and sensitivity of the speaker in the poem. Or is that, too, relative?

What happens to sound: Where does it go? Is sound forever or does sound end? George Santayana, (1863-1952) another twentieth century poet, provides his answer in the poem, "Odes."

> Gathering the echoes of forgotten wisdom,
> And mastered by a proud, adventurous purpose,
> Columbus sought the golden shores of India
> - from "Odes" -
> - George Santayana -

He views the printed word as a continuation of sound -- the recorded sounds of thoughts continue. Sound does not end. The sounds, the words, the wisdom, all continue. But a few civilizations, such as the Andean Incas, had no written language. They left no words; their sounds ended.

John Greenleaf Whittier (1807-1892), poet and journalist, proposed a more concrete characteristic of sound. He purported that once the physical energy is released as sound, it continues to exist -- "Its spectre lingers round." He used this idea in his poem entitled, "An Autograph." Whittier constructed a parallel between the continuing presence of sound and his yearning that remembrances of his existence will outlast him and, "Leave some faint echo still."

The seventh stanza is the pivotal point in the poem. The first stanzas forlornly lament the apparant inevitablility of one's passing from memory. The last stanzas are much more bouyant because he believes that life will, "Leave some faint echo still."

> I write my name as one,
> On sands by waves o'errun
> On winter's frosted pane,
> Traces a record vain.
>
> Oblivion's blankness claims
> Wiser and better name
> And well my own may pass

As from the strand or glass.

Wash on, O waves of time!
Melt, noons, the frosty rime!
Welcome the shadow vast,
The silence that shall last!

When I and all who know
And love me vanish so,
What harm to them or me
Will the lost memory be?

If any words of mine,
Through right of life divine,
Remain, what matters it
Whose hand the message writ?

Why should the "crowner's quest"
Sit on my worst or best?
Why should the showman claim
The poor ghost of my name?

Yet, as when dies a sound
Its spectre lingers round,
Haply my spent life will
Leave some faint echo still.

A whisper giving breath
Of praise or blame to death,
Soothing or saddening such
As loved the living much.

Therefore with yearnings vain
And fond I still would fain
A kindly judgement seek,
A tender thought bespeak.

And, while my words are read,
Let this at least be said:
"Whate'er his life defeatures,
He loved his fellow-creatures.
               - "An Autograph" -
               - John Greenleaf Whittier -

The focus throughout the book has been upon poets and their poetry. We shall digress briefly at this point to examine one short excerpt from the prose of the eighteenth century New England poet, Henry David Thoreau (1817-1862). The reason is two-fold.

First, although the passage is prose, the lyrical talents of the poet permeate the writing. It could, in fact, be considered open verse. The second reason for its inclusion is because of the sophistication of his use of the properties of sound, and of the subtleties of meaning that can escape us if it is hastily read.

If a man does not keep pace with his companions,
perhaps it is because he hears a different drummer. Let
him keep step to the music he hears however measured
or far away.
       - *Walden* -
       - Henry David Thoreau -

The simplest interpretation is made from the literal reading of the first sentence. If one appears to be out-of-step with the others, then logically each is responding to something different. But different responses are produced from different stimuli or circumstances. Thoreau moderates the obvious when he says, "perhaps . . . he hears." This leaves open the possibility that the same stimulus can be perceived and interpreted quite differently. The second sentence builds upon this idea that only one tune may be playing. The differences among the listeners as to what they hear is because of their perceptions and position within the context. Thoreau bids the listener to, "keep step to the music," to be true to what his values and ideals say to him.

We are aware within the second sentence of the use of the more subtle principle of sound when he says, "however measured or far away." If we visualize a parade approaching us from some distance, our first reaction is dismay that the marchers are out-of-step with their music; they bob antagonistically to the beat. Simultaneously, the distant marchers see us tapping our feet, and they are appalled that we are out-of-step with their music. Each is certain that the other is out-of-step.

Of course the reason for the apparent lack of synchrony is because of the relative differences between the speed of light and the speed of sound. Light, traveling at the velocity of 186,000 miles per second, travels the quarter-mile between us and the marchers in a mere fraction of a second. We see their actions instantaneously. Sound, traveling at a velocity of 1100 feet per second, travels the quarter-mile in somewhat more than a second. We see the movement instantly, but the sound of the beat is delayed.

Thoreau calls upon us to maintain our pace, to keep to our ideals, because as the distance that separates us decreases, the beat of the music and our footfalls will slowly become synchronized. He is also warning that we should not judge too quickly the lack of harmony. Disharmony may be due to different music, but it may more likely be due to our distance from each other. The message becomes even clearer when we substitute the word understanding for distance. Disharmony decreases as the distance becomes closer; disharmony decreases as the understanding becomes closer.

An appropriate conclusion to the chapter on light and sound is the sonnet entitled "Genesis" by A. M. Sullivan (1896-1980). In it we have an excellent example of the blending of the two physical phenomena. Each complements the other and builds to a unified praise of the Creation.

> The light is music, and the first note breaks
> The shuttered silence like a bird in flight;

This is not morning but the end of night,
The long night when the urge for being aches
In Time's slow womb, and a windy rhythm shakes
The shadows from the sun, and from the height
And breadth of Heaven fades the sullen blight
Of emptiness and God in joy awakes.

The layers of aeons crumble from the dark
Stirred by the anthem of angelic strings
And Order comes with a lordly whirr of wings
And worlds are made in the expanding arc
Of splendor till a last note finds its mark
And Man spits out the primal clay and sings.
              - "Genesis" -
              - A. M. Sullivan -

CHAPTER VII   EARTH AND SKY

> To know truth partially is to distort
> the universe.
> -from *Adventures of Ideas*-
> -Alfred North Whitehead-

In the beginning

On October 26, at 9:00 A.M. in the year 4004 B.C., the earth began.
    To those of us today who think of the earth's age in terms of billions of years, we find it difficult to take seriously Archbishop James Ussher's (1581-1651) proclamation that 4004 B.C. was the "beginning," and we are tempted to treat his undertaking lightly. Yet it was a significant scholarly achievement that resulted from a laborious tracing of the Biblical geneologies and a correlating of them with the various changes in the calendar. If we put ourselves into the context of the seventeenth century, the date had religious and philosophical implications far more important than any scientific inaccuracy.

The literal minded of the church fathers recognized no distinction between historical and mythical time. They viewed the geneologies of Genesis as a complete and accurate chronicle of the time since the first day of Creation. The beginning could thus be dated accurately and exactly in years -- and as Ussher showed, in months, days, and even minutes.

It is important to know that Ussher was not the first to make such a calculation. Theophilus of Antioch (115-181 A.D.) probably deserves credit for being the first to undertake a serious historically based determination of the earth's beginning. Using histories and manuscripts, he traced backwards from the beginning of the Christian era and ascertained 5529 B.C. as the date of Creation. Although he recognized a certain impreciseness inherent in his work, he defended himself in his remarks to those predecessors who had proposed the earth's age to be millions of years old:

> For even if a chronological error has been committed by us, of, e.g., 50 or 100, or even 200 years, yet (it is) not of thousands and tens of thousands, as Plato and Apollonius and other mendacious authors have hitherto written.[1]

The next significant contribution following Theophilus' work was that of Julius Africanus (200-250 A.D.) whose calculated date for the earth's creation was 5500 B.C. This date was not so much different from Theophilus' date as was his reasoning. His theory was that all history must comprise the cosmic week because it took God a week for his Creation. And furthermore, each day of that week lasted a thousand years. The length of the cosmic day came to him directly from the scripture:

> 4 For a thousand years are in thy sight as but
> yesterday when it is past, as a watch in the night.
> - Psalm 90 -

> 8 But, beloved, be not ignorant of this one thing,
> that one day is with the Lord as a thousand years,
> and a thousand years as one day.
> - II Peter 3 -

As time progressed, other Scriptural chronologists developed the Creation date. Although each differed slightly depending upon the variations in the texts they used, they all agreed that the age of the earth was only slightly more than the history of man. And for that reason they were all restricted to a period of from five to six thousand years.

Eschatologists, whose theology concerned the final events in the history of the world, also had a profound influence on the Christian time scale. These apocalyptical writers accepted totally the Biblical account of Creation as found in Genesis, but they also projected beyond it to the inevitable day of destruction and judgement. They regarded Creation as a prophecy of the ages of the world. If the world were created in six periods of a thousand years, then obviously, wrote Lacantius in *The Divine Institute*, it would last for six periods.

> . . . . we, whom the Holy Scriptures instruct to the knowledge of the truth, know the beginning and end of the world. . . . God completed the world and this admirable work of nature in the space of six days, as is contained in the secrets of Holy Scripture, and consecrated the seventh day, on which He had rested from His works. . . .
>
> Therefore, since all the works of God were completed in six days, the world must continue in its present state through six ages, that is, six thousand years. For the great day of God is limited by a circle of a thousand years, as the prophet shows, who says, "In Thy sight, O Lord, a thousand years are as one day."[2]

Martin Luther (1483-1546) wrote that the history of world was to be expressed in six ages, the age of Adam, Noah, Abraham, David, Christ, and the Pope. And he added that the Pope would not complete his millenium. "It is my firm belief that the angels are getting ready, putting on their armor, and girding their swords about them, for the last day is already breaking . . . . The world will perish shortly."[3]

By the middle of the seventeenth century the Aristotelian concept of the antiquity of the earth was almost completely forgotten and ignored. The Christian scriptural view, which envisioned an omnipotent God creating the world in 4004 B.C., and the Jewish calculation of 3760 years before the common era, derived their holy conclusions totally from Genesis. Man and his world were approaching their end. Laments regarding old age and sickness, be they earth's or man's, were not mere rhetoric, but were based upon unquestionable fact which was validated by the Bible itself. Rosalind in Shakespeare's, *As You Like It*, sighed: "The poor world is almost six thousand years old." Sir Thomas Browne, the seventeenth century essayist, likening the aging world to his own advancement, worte: "The World grows near its end . . . . The last and general fever may as naturally destroy it before six thousand, as me before forty."[4]

Still, although overshadowed for centuries by the Biblical determinations, the ancient idea about the antiquity of the earth was not content to remain in the darkness. The Pre-Christian era had birthed the Chaldean, Babylonian, and Persian hypotheses of immense cosmic cycles. The Indian cycle, or Mahayuga, for example lasted 12,000 years in a sequence of creation-destruction-creation, with each of these divine years equal to 360 earth years. A cosmic cycle was in excess of four million years, yet constituted only one day in the life of Brahma. And the Platonic Year, known also as the magnus annus, had its own 36,000 year cycle.

The revitalization of the idea of the earth's immense age was an inevitable result of the new science beginning in the seventeenth and eighteenth centuries. Copernicus and his *Revolutions of the Celestial Spheres*, Kepler and his laws of planetary motion, and Lippershey's telescope eliminated the concept of the heavens being a blue ceiling over the unique earth, and replaced it with the reality of a seemingly limitless universe of which earth was only one speck -- and an insignificant speck at that. Furthermore, the Newtonian universe led thoughts away from the idea of a world created fully-formed in an instant of time.

As part of the rising wave of challenges to the existing idea, Comte de Buffon (1707-1788) utilized a new technique from the arsenal of science to estimate the age of the earth. Relying upon the principles of thermodynamics he assumed a molten condition for the earth's initial existence and applied a cooling-rate equation to calculate how long it would take for the earth's present temperature to be reached. His conclusion of 75,000 years as the earth's age was very much less than the age we accept today, but the significance of his effort was that science was being used. The seeker had moved from the library to the laboratory, and his conclusion had broken sharply with historical time and thus freed the age of the earth from its bondage to the age of man; it spoke to an antiquity far beyond that of the Biblical chronologists.

Other scientists began using still different methods of estimating the earth's age. James Hutton (1726-1797) saw no need to invoke Noah's Flood or other catastrophes to explain the earth's topography and physical degradation from its original perfection. His thesis, uniformitarianism, was simply that the normal agents of geological change could, if allowed adequate time, account for all of the changes that have shaped the earth. Although he carefully avoided the projection of a precise age of the earth, he did speak of it in terms of fossils and the times required for transformations as being so incomprehensively long as to defy calculation. It was left to Charles Darwin (1809-1822), the evolutionist, to quantify the age to 300 million years based upon an interpolation of erosion and deposition.

At first the scientists' new attempts to redefine the earth's age were met with widespread skepticism. William Cowper mocked their efforts in his poem "The Task." Quite simply we see him equating the misguided earth scientists with the inept writers who purport to be exposing truth but who waste their time, and their lives, truly making mountains from molehills. What audacity it was to challenge the Holy Word as was given to Moses!

> Some write a narrative of wars and feats
> Of heroes little known, and call the rant
> An history. Describe the man, of whom
> His own coevals took but little note,
> And paint his person, character and views,
> As they had known him from his mother's womb.
> They disentangle from the puzzled skein
> In which obscurity has wrapp'd them up,
> The threads of politic and shrewd design,
> That ran through all his purposes, and charge
> His mind with meanings that he never had,
> Or having, kept conceal'd. Some drill and bore
> The solid earth, and from the strata there
> Extract a register, by which we learn
> That he who made it and reveal'd its date
> To Moses, was mistaken in its age.
> Some more acute and more industrious still
> Contrive creation. Travel nature up
> To the sharp peak of her sublimest height,
> And tell us whence the stars. Why some are fixt,
> And planetary some. What gave them first
> Rotation, from what fountain flow'd their light.
> Great contest follows, and much learned dust
> Involves the combatants, each claiming truth,
> And truth disclaiming both. And thus they spend
> The little wick of life's poor shallow lamp,
> In playing tricks with nature, giving laws
> To distant worlds and trifling in their own.
> Is't not a pity now that tickling rheums
> Should ever teaze the lungs and blear the sight
> Of oracles like these? Great pity too,
> That having wielded th' elements, and built

> A thousand systems, each in his own way,
> They should go out in fume and be forgot?
> Ah! what is life thus spent? and what are they
> But frantic who thus spend it? all for smoke --
> Eternity for bubbles, proves at last
> A senseless bargain. When I see such games
> Play'd by the creatures of a pow'r who swears
> That he will judge the earth, and call the fool
> To a sharp reck'ning that has liv'd in vain,
> And when I weigh this seeming wisdom well,
> And prove it in th'infallible result
> So hollow and so false -- I feel my heart
> Dissolve in pity, and account the learn'd,
> If this be learning, most of all deceived.
>   - from "The Task: Book III" -
>   - William Cowper -

The theological implications of the renewed interest in the non-Biblical dating of the earth's age were more significant than we might at first think. According to Genesis, the world and all of the inhabitants and forms were created instantly, although sequentially, by divine fiat. Orthodoxy feared that if geological ages, scientifically deduced, were admitted, then the literal words of Genesis and all of the doctrine based upon it would be subject to disputation. The impact could be catastrophic if in fact, as Cowper had written, "That he who made it and reveal'd its date / To Moses, was mistaken in its age."

Astronomy

By the early twentieth century the telescope and spectroscope were permitting astronomers to make new inferences about the universe. The anonomously written contemporary version of a familiar nursery rhyme attests to the revelations gained from these instruments.

> Twinkle, twinkle little star
>   I don't wonder what you are.
> For by the spectroscopic ken
>   I know that you are hydrogen.[5]

Scientist could now determine celestial motion precisely-- that is, the speed and direction of movement of far off stars-- and with a high degree of reliability. Contributing to this capability was the discovery that the Doppler Effect, thought to be peculiar to sound only, had a parallel with light. Christian Doppler (1803-1853) had determined that as a noisy object traveled toward the listener, the pitch was high; when the same object traveled away from the listener the pitch was lower.

Astronomers' obervations of light seemed to show a similar pattern. If light was approaching the viewer, the light was of a shorter wave length and its spectrum shifted toward the blue. As the light moved away from the viewer, the wavelength was longer and shifted toward the red. These observations were made possible by spectroscopically examining Newton's gift -- the visible spectrum. Not only were astronomers able to determine the direction the light object was traveling, but also they were able to determine its speed by measuring the amount of shift toward the red or the blue end of the visible light spectrum.

In the 1910s and 1920s astronomers had come to the conclusion from their observations that almost all of the stars were receding, and at varying velocities. It was left to Edwin Powell Hubble in 1929 to decipher the relationship and to express it in the simple terms that became known as Hubble's Law. He deduced that velocity was proportionate to distance. The farther out a star was located, the faster it was traveling away from the viewer. The implication of Hubble's Law was that the universe had a limited size! If Einstein's theory was correct, that an object's mass becomes infinite when it attains the velocity of light, then when a receding star reached that velocity, its mass would become infinite and the end of the universe would have been reached. Hubble calculated the radius of the universe to be 12 billion light years.

(If light travels at a velocity of 186,000 miles per second, Hubble's Radius calculates to be 70,000,000,000,000,000,000,000 miles or 7×10^{22}.)

In spite of Hubble and, to some degree, Einstein, the idea of the limited universe did not predominate within the scientific community. The theory of the limitless, expanding universe continued to gain support even though bothersome problems persisted. If the univese were expanding and the stars and galaxies were all proceeding outward at some incredible velocity, then the universe would ultimately become "empty." New vacuum would be created as the mass moved outward. In 1948 Fred Hoyle proposed the Steady State Theory as an answer. He theorized that the new matter is being created continuously, but also that it is being created at the exact rate to compensate for the vacating matter. His theory implied an infinite universe, infinite in age, infinite in dimension, and infinite in its production of matter.

We recognize immediately that the Steady State Theory contradicted the scientific and religious beliefs that depict an instant of Creation. With no instant of Creation, Genesis would need to be rewritten from "God created..." to "God began creating..." The Steady State Theory described a universe with no locus of beginning, an infinite past, and an infinite future.

George Gamov (1904-1968) was a strong proponent of the concept of the expanding universe, but not a proponent of the Steady State. He entered into the controversy with a big bang -- that is -- the Big Bang Theory!

Gamov visualized the stars and galaxies flying apart like shrapnel from a bomb. His theory defined a singular instant of Creation -- The Big Bang -- before which our universe did not exist as we know it. Furthermore, his theory allowed for the calculation of the place and time of the Creation. If the galaxies were moving outward at a certain velocity, then one would need only to trace backwards along each one's path until the point was reached where all of the galaxies converged. That would be the place of Creation. By knowing the velocity, one could readily calculate the time required to cover the distance traveled from the locus of the

Big Bang to the current celestial position. Gamov placed the date of Creation as 20 billion years ago. But still, he volunteered no explanation about how the matter itself was first created!

Look now at the poem by the mid-twentieth century poet May Swenson. Consider how the poet communicated three theories of the universe.

1.
At moment X
the universe began.
It began at point X.
Since then,
through the Hole in a Nozzle,
stars have spewed. An
inexhaustible gush
populates the void forever.

2.
The universe was there
before time ran.
A grain
slipped in the glass:
the past began.
The Container
of the Stars expands;
the sand
of matter multiplies forever.

3.
From zero radius
to a certain span,
the universe, a Large Lung
specked with stars,
inhales time
until, turgid, it can
hold no more,

and collapses. Then
space breathes, and inhales again,
and breathes again: Forever.
- "3 Models of the Universe" -
- May Swenson -

Swenson presented another description of the universe in her poem entitled "Cardinal Ideograms." What number would one choose to represent the universe? She chose the number 8. The eight has the shape of an hourglass -- representing finite time, but with endless repetitions and cycles. It has no beginning; it has no ending. The 8 is discrete in quantity, constant in value, yet, when laid on its side it becomes the symbol for infinity.

8   The universe in diagram:
A cosmic hourglass.
(Note enigmatic shape,
absence of any valve of origin,
how end overlaps beginning.)
Unknotted like a shoelace
and whipped back and forth,
can serve as model of time.
- from "Cardinal Ideograms" -
- May Swenson -

But speculations about the universe are not the exclusive domain of the twentieth century poet. Two thousand years ago the Roman poet, Lucretius, in his lengthy poem, *De Rerum Natura*, explained to the reader the certainty of the limitless universe. Notice how this idea evolved from what one would describe as common experience.

I have already taught you
That matter's basic elements are solid,
Completely so, and that they fly through time
Invincible, indestructible for ever.

Now let's work out whether there's any limit
To their sum total; study, likewise, void,
Space, emptiness, area where all things move.
Does this have finite limits or does it reach
Unmeasurable in deep wide boundlessness?
The universe is limitless, unbounded
In any of its areas; otherwise
It would have to have an end somewhere, but no --
Nothing, it seems, can possibly have an end
Without there being something out beyond it,
Beyond perception's range. We must admit
There can be nothing beyond the sum of things,
Therefore that sum is infinite, limitless.
It makes no difference where you stand, your center
Permits of no circumference around it.
Assume, though, for a moment, that all space
Is definitely limited, what happens
If somebody runs to its furthest rim, and rifles
A javelin outward? Will it keep on going,
Full force, or do you think something can stop it?
Here's a dilemma that you can't escape!
You have to grant an infinite universe
For either there's matter there, to stop our spear,
Or space through which it keeps on flying. Right?
So it wasn't flung from any boundary line.
I will keep after you with this argument, ask you,
No matter where you set the outermost limit,
What happens to the javelin after that?
The answer is that final boundary line
Is nowhere in existence, there will always
Be plenty of room beyond for the spear's flight.
Before our eyes, thing seems to limit thing,
Air bounds the hills, and forests border air,
Earth sea, sea earth, but add them up, and nothing
Limits the sum.

      - from *De Rerum Natura* I, 951-989 -
      - Lucretius -

Common experience, however, does have its traps. The universe, writes Lucretius, cannot be bounded like the walls of a box. If it were, all matter would fall to the bottom. His common experience taught him that all things fall to the bottom. It was fifteen hundred years later that Newton showed the world that gravity is the attractive force operating between any two or more bodies. Attraction -- falling -- can be either downward or upward. Gravity is not a unidirectional force in the universe.

>   Besides, if all the space
> Of all the universe were fixed, enclosed
> By definite bounds, by this time all the mass,
> The weight, of matter would have run together
> From all sides to the bottom, tending downward
> From the sheer force of weight, so there could be
> No room for action under heaven's roof,
> No heaven, for that matter, and no sun,
> Since all material would be heaped so high
> From its long subsidence through endless time.
> But as it is, no rest is ever given
> To the atoms' rainfall; there's no pit, far down,
> To be their pool, their ultimate resting-place,
> All things keep on, in everlasting motion,
> Out of the infinite come the particles
> Speeding above, below, in endless dance.
> By nature space is deep and space is boundless,
> So that bright shafts of lightning could not cross it,
> Given eternal time, nor could not cross it,
> The area before their onward course.
> There is too much space, all here and there, around them,
> No limit to that infinite domain.
>   -from *De Rerum Natura* I, 990-1007-
>   -Lucretius-

Departing from the absolute certainty of Lucretius two

thousand years ago, consider now the uncertainty of Robert Frost in the twentieth century. He used his poem, "Accidentally on Purpose," to describe his fundamental uncertainty about what underlies the universe and man's existence. He acknowledges the universe is "but the Thing of things,\ The thing but balls all going round in rings," but he questions the belief that "the omnibus\ Had no real purpose till it got to us." He goes further to admit by implication that he really does not know the truth and so he selects his own version which may well be "the very worst." Nevertheless, his real attitude seems to be indifference, whereby he will just leave the truth to "scientific wits" to figure out. As Professor Robert Thompson writes: "Frost has compensated his realization of the accidental nature of the universe with a simple statement of disbelief . . . . the disbelief provides the basis for an uncertainty which signalizes the poet's move from romanticism towards existentialism."[6]

> The Universe is but the Thing of things,
> The things but balls all going round in rings.
> Some of them mighty huge, some mighty tiny,
> All of them radiant and mighty shiny.
>
> They mean to tell us all was rolling blind
> Till accidentally it hit on mind
> In an albino monkey in a jungle,
> And even then it had to grope and bungle,
>
> Till Darwin came to earth upon a year
> To show the evolution how to steer.
> They mean to tell us, though, the Omnibus
> Had no real purpose till it got to us.
>
> Never believe it. At the very worst
> It must have had the purpose from the first
> To produce purpose as the fitter bred:
> We were just purpose coming to a head.

Whose purpose was it? His or Hers or Its?
Let's leave that to the scientific wits.
Grant me intention, purpose, and design--
That's near enough for me to the Divine.

And yet for all this help of head and brain
How happily instinctive we remain,
Our best guide upward further to the light,
Passionate preference such as love at sight.
-"Accidentally on Purpose"-
-Robert Frost-

While we are considering the universe, we should pause for a moment to consider the moon, stars, and planets and the mystical force they exerted upon the thinking of man -- astrology was that force.

Astrology

Historians believe that astrology probably began in the pre-Babylonian period by the Chaldeans, and was based upon the study of the sun, the moon, and the five planets known at the time. The seemingly mystical behavior of these celestial bodies led ancient man to the conclusion that clues to the future were hidden among their patterns. As time progressed, soothsayers and astrologers flourished, and their importance increased as their capacities to predict the future became more relied upon by the rich and the poor, by the powerful and the exploited.

Yet, as the influence of the astrologer increased, so did the fear of him increase. In A.D. 11, the Roman emperor Augustus was forced to issue an edict limiting the topics about which clients were allowed to consult them. The general faith in astrology was so great that political predictions about the government and its participants constituted prima facie evidence of treason. Illegal astrological inquiries included those which were related to the emperor's health, to high matters of state, to one's own future well-being, to the well-being of one's family, and to the well-being of masters when inquired by slaves. Punishments were flogging for the first offenders, and repeating violators were sent to the mines or an island exile.[7]

The deterioration of the acceptability of astrology by officialdom continued through the first several centuries. By the fourth century the Christian rulers had outlawed astrologers and the practice of astrology throughout the Roman Empire. These restrictions, however, were based on religious grounds not political ones as in the first century. Even so, a type of imperial hypocrisy was exercised. Astrology was considered to be the most nearly infallible method of divination. Any emperor, therefore, would feel duty bound to avail himself of the art in order to discover future dangers to the Empire (and of course to himself).[8]

The Renaissance was a period of strong belief in astrology, thus we are not surprised to see poets of the era -- John Lyly, Sir Thomas Browne, and William Shakespeare -- incorporating the elements of astrology into their works. Shakespeare, for example, was neither a scientist nor an astrologer; he was a poet and a playwright. He created characters and circumstances that his public wanted to see and hear, and with which they could relate. Whether or not the astrological references were reflections of his beliefs is less important than the fact that they were obviously compatible with the beliefs of the audiences of the time.

In the book *Tamburlaine's Malady*,[9] Professor Johnstone Parr of the University of Alabama wrote:

> The use of astrology that looms largest in Shakespear's dramas is the free utilization of astral philosophy in the creation of some especially artistic and beautiful lines. Innumerable astrological passages in the plays are so composed as to make them particularly striking and dramatically effective. In such lines as the following we see the master artist taking common astral tenets and -

apparently not caring whether the reluctant astrology is "scientifically" correct or not - remoulding his raw materials into something effective and artistic:

Prospero: I find my zenith doth depend upon
A most auspicious star, whose influence
If now I count not, but omit, my fortune
Will ever after droop.
-*The Tempest*, I.ii. 181-184-

Othello: It is the very error of the moon:
She comes more nearer earth than she was wont,
And makes men mad.
-*Othello*, V.ii. 109-111-

Ulysses: The heavens themselves, the planets, and this centre,
Observe degree, priority, and place,
Insisture, course, proportion, season, form,
Office and custom, in all line of order:
And therefore is the glorious planet Sol
In noble eminence enthroned and spher'd
Amidst the other; whose medicinable eye
Corrects the ill aspect of planets evil,
And posts, like the commandment of a king,
Sans check to good and bad: but when planets,
In evil mixture, to disorder wander,
What raging of the sea! shaking of the earth!
Commotion in the winds! frights, changes, horrors,
Divert and crack, rend and deracinate
-*Trolius and Cressida*, I.ii. 85-102-

Cassius: The fault, dear Brutus, is not in our stars,
But in ourselves, that we are underlings.
-*Julius Caesar*, I.ii. 140-141-

Calpurnia: When beggars die, there are no comets seen;
The heavens themselves blaze forth the
death of princes.
-*Julius Caesar*, II.ii. 30-31-

Richard: Lo, at their births good stars were opposite!
-*Richard III*, IV.iv.212-

Helena: Monsieur Parolles, you were born under a charitable star.
Parolles: Under Mars, I.
Helena: I especially think under Mars.
Parolles: Why under Mars?
Helena: The wars have kept you so under, that you must needs be born under Mars.
Parolles: When he was predominant.
Helena: When he was retrograde, I think rather.
Parolles: Why think you so?
Helena: You go so much backward when you fight.
-*All's Well*, I.i. 206-219-

Elizabethans strongly believed that the heavenly bodies influenced, at least to some degree, the events of one's life, everyday happenings, and the behaviors and actions of people. Significant among the beliefs was that which characterized the planets as animate beings. Numerous painting and woodcuts illustrated the personification of "planets" ie Sol, Venus, Saturn, Luna et al. Some writers combined this personification with the idea of the inherent properties of the planets which were firmly believed to govern those persons who fell under their shadow. A idea of the inherent properties of the planets which were firmly believed to govern those persons who fell under their shadow. A reflection of this thought is demonstrated by the Elizabethan poet-drammatist John Lyly (1554-1606) in his play *The Woman in the Moone*.

The plot is simple. Pandora is the perfect woman; the living perfection of the gods' and goddesses' creation. The seven planets, however, chagrined because they were not involved in her creation, decide to destroy this perfection. Each planet, therefore, participants in the ruination by exerting its own unique influence on her. Lyly personifies each planet and has it speak of its own malevolence.[10]

> Satrun speaks:
> I shall instill such melancholy moode,
> As by corrupting of her purest blood,
> Shall first with sullen sorrows clowde her braine,
> And then surround her heart with forward care:
> She shall be sick with passions of the heart,
> Selfwild, and toungtide, but full fraught with teares.
> - *The Woman in the Moone* I.i.144-149 -

> Jupiter speaks:
> Now Jupiter shall rule Pandoraes thoughts,
> And fill her with Ambition and Disdaine;
> I will inforse my influence to the worst,
> Least other Planets blame my regiment.
> - II.i.2-5 -

> Mars speaks:
> Now bloody Mars begins to play his part,
> Ile worke such warre within Pandoraes breast,
> That after all her churlishness and pride
> She shall become a vixen Martialist.
> - II.i.177-180 -

> Mars hath inforst Pandora 'gainst her kinde,
> To manage armes and quarrel with her friends:
> And thus I leave her, all incest with yre:
> Let Sol cool that which I have set on frye.
> - II.i.236-239 -

> Luna speaks:
> Now other planets influence is done,
> To Cynthia, lowest of the erring starres,
> Is beautious Pandora given in charge.
> And as I am, so shall Pandora be,
> New Fangled, Fyckle, slothful, foolish, mad.
> - V.i.1-6 -

The poet, with his sometimes unique perspective, can frequently turn ideas to his own benefit. The seventeenth century poet, George Wither, found a way to release himself from the pangs of guilt over his transgressions as a mortal man. He asked, simply, why he should be blamed for sin when it was the planets' fault.

> Some say (and many men do these commend)
> That all our deeds and Fortunes do depend
> Upon the motions of celestial Spheres,
> And on the constellations of the Stars.
> If this were true, the Stars alone have been
> Prime cause of all that's good, and of all sin.
> And 'twere (me thinks) injustice to condemn,
> Or give rewards to any, but to them.
> For if they made me sin, why for that ill
> Should I be damn'd my goodness, why should I
> Be glorified for their Piety?
> And if they neither good nor ill constrain.
> Why then should we of Destiny complain?
> For if it be (as 'tis) absurd to say
> The stars enforce us (since they still obey
> Their just Commander) 'twere absurder far

> To say, or think, that God's Decree it were
> Which did necessitate the very same,
> For which we think the stars might merit blame.
> He made the stars to be an aid unto us,
> Not (as is fondly dream'd) to help undo us:
> (Much less, without our fault, to ruinate,
> By doom of irrecoverable Fate)
> And if our good Endeavours use we will,
> Those glorious creatures will be helpful still
> In all our honest ways: For they do stand
> To help not hinder us, in God's command;
> And he not only rules them by his pow'rs,
> But makes their Glory servant unto ours.
>   - "God, by their Names, the Stars doth call" -
>   - George Wither -

The importance of astrology probably peaked in the fourteenth and fifteenth centuries, and began to wane as the influence of Copernicus and Galileo replaced the mysticism of the heavens with a scientific base. Yet, is astrology a myth or science? -- hoax or truth? Its believers unhesitatingly accept astrology as an imperative to survival -- the ultimate illumination of the darkness of the unknown. Its disbelievers unhesitatingly characterize astrology as an afront to scientific knowledge -- the ultimate hoax upon the ignorant. So serious do they view its threat that in 1975 a formal statement entitled "Objections to Astrology"[11] was issued by 192 scientists from around the world, including 19 Nobel prize winners.

> ...It is simply a mistake to imagine that the forces exerted by the stars and planets at the moment of birth can in any way shape our futures. Neither is it true that the position of heavenly bodies make certain days or periods more favorable to particular kinds of action, or that the sign under which one was born determines one's compatibility or incompatibility with other people... We

believe that the time has come to challenge directly and forcefully the pretentious claims of astrological charlatans.

Still today, many "good" newspapers continue to feature their resident soothsayer's predition of what is going to happen today and tomorrow, and how our fortunes and lovelife will fare.

Geology

As the newest of the sciences, geology's impact in the nineteenth century was unexpectedly influential to the extent that the study of the earth was largely responsible for the spread of the idea of evolution. The stratefication of the earth's surface, and the existence of fossils, provided a manuscript of the earth's history as clear to the scientists as the early scrolls had been to the Scriptural scholars. The geologist-poet of the late nineteenth century, William Pallister, described the earth scientists' work:

>The story of the earth's evolution, told
>In rocks which held the pages stratified:
>In granites torn by glacier and tide,
>In limestone, coal-bed, river-clay and mould;
>The story of the past, now blurred, now bold,
>Which nature tells in signs on every side,
>The story of the species, all allied.
>Men patiently translate these signs; behold!
>Our lakes are salt, three times their present size!
>Old ice-caps melt, new life of every kind
>Appears and leaves wierd fossils when it dies,
>A rising measure of the larger mind.
>Men see the method, how the world force acts
>To build the wonder of the present facts.
>   - "Geology" -
>   - William Pallister -

Logically the geologists were among the last of the scientists to be targeted by the romantic poets. William Wordsworth poured contemptuous pity on the breed of man who would invade the quiet hillside, hammer in hand, chipping and scarring the moss-covered rocks -- mutilating the pathways along which the poets chose to wander and meditate.

> Nor is that Fellow-wanderer, so deem I,
> Less to be envied, (you may trace him oft
> By scars which his activity has left
> Beside our roads and pathways, though,
>   thank Heaven!
> This covert nook reports not of his hand)
> He who with pocket-hammer smites the edge
> Of luckless rock or prominent stone, disguised
> In weather-stains or crusted o'er by Nature
> With her first growths, detaching by the stroke
> A chip or splinter -- to resolve his doubts;
> And, with that ready answer satisfied.
> The substances classes by some barbarous name,
> And hurries on; or from the fragments picks
> His specimen, if but haply interveined
> With sparkling mineral, or should crystal cube
> Lurk in its cells -- and thinks himself enriched.
> Wealthier, and doubtless wiser, than before!
>   - "The Excursion" Book III -
>   - William Wordsworth -

Not only was the poet disturbed by the intrusion of the geologist into his tranquil domain, but also the theologist-philosopher was disturbed as his domain, the Great Chain of Being, was similarly intruded upon by the earth scientist. The belief in the Chain of Being, that each member of the Creation had a fixed and predetermined place, was being eroded by proof of the disappearance and mutation of species. And man, theologically regarded as a special creation set apart from the animals, suddenly found himself to be related to them.

John Ruskin (1819-1900), trying to convince himself that geology and traditional religion could be reconciled, wrote of his frustration. His faith was "being beaten into mere gold leaf, and flutters in weak rags from the letter of its old forms .... If only the Geologists would leave me alone, I could do very well, but those dreadful hammers!" he lamented, "I hear the clink of them at the end of every cadence of the Bible verse..."[12]

It is appropriate to digress for a moment to show again the impact that Isaac Newton had upon science and language. The new science that Newton described in his book, *Optics*, gave the poets an enhanced awareness of color and the principles of light. As a result, we find some of them writing verses to display this awareness of the new knowledge, and frequently they combined the new science of light with the new science that was geology. Henry Brooke (1703-1783) included a roll-call of colors in his description of gems.

> Through sparkling gems the plastic artists play,
> And petrify the light's embody'd ray;
> Now kindle the carbuncle's ruddy flame,
> Now gild the chrysolite's transparent beam;
> Infuse the sapphire's subterraneous sky,
> And tinge the topaz with a saffron dye;
> With virgin blush within the ruby glow,
> And o'er the jasper paint the show'ry bow.
> - from "Universal Beauty" -
> - Henry Brooke -

William Thompson, the eighteenth century poet, combined scripture with Newtonian science, insisting that the laws of refraction held in Heaven as on earth.[13]

> A mirror spreads its many-colour'd round,
> Mosaic-work, inlaid by hands divine
> In glist'ring rows, illuminating each,

> Each shading: beryl, topaz, chalcedon,
> Em'rald and amethyst. Whatever hues
> The light reflects, celestial quarries yield,
>
> Or melt into the vernant-showry bow,
> Profusive, vary here in mingling beams.
> - from "Sickness" -
> - William Thompson -

Earth Sciences

We find that poets have developed strong and meaningful analogies between various phenomena of the earth and conditions of man. The contemporary poet, Robert Morgan, has struck a parallel between the unseen scissions that separate mankind. The cracks, often of unknown origin, have aged and been covered with a veneer of mantle to the extent that from the surface they no longer appear to exist. But the split remains and "divides the continent against itself." Is he saying, or is he asking, if the same "grudge of kinship" exists with mankind?

> The crack in our hearth of land
> runs almost imperceptible under hills
> and riverbottoms, slicing ranges.
> So much folding and drift and erosion
> have covered the flaw
> the terrain appears seamless
> except for the belt of crushed rock
> strewn across the drainages.
> But the split, deeper than any spring or cellar,
> older even than the rise of mountains,
> touches the fire of the orignial
> schism, and reaching beyond highland,
> piedmont and coastal plain, divides
> the continent against itself

    and builds the grudge of kinship
    under quiet blue slopes.
                     - "Brevard Fault" -
                       - Robert Morgan -

    In another poem, A. M. Sullivan (1896-1980) describes the seismograph as "the pain recorder of the terrestrial ache." Quakes, vulcanism, and "the creaking joints in the fevered mountains" are measured by this device that is sensitized to the twitches of the earth's skin. And yet, the real earth-pains -- war and hate -- go undetected.

    The earth twitches its skin
    Like a goaded elephant
    And jars the electric pin
    Which measures the irritant.

    This is the pain recorder
    Of the terrestrial ache
    And sea disorder
    By which men take
    Heed of pressure points
    By volcanic fountains,
    And listen to creaking joints
    In the fevered mountains

    It records earth's wounds
    But never the patter
    Of marching feet, nor sounds
    Of iron throats that scatter
    Hate on her crust.
                      - "Seismograph" -
                      - A. M. Sullivan -

    Aside from quakes and vulcanism, other phenomena of the earth's crust have been used to give a special meaning to -- or in some cases, to confound the meaning within -- the poem. A particular example of this type is the poem, "As I Walked Out One Evening" by W. H. Auden (1907-1973). Before reading it, however, one must consider the phenomena known as continental drift.

In the early nineteenth century, cartographers and earth scientists became increasingly intrigued by the amazing relatedness of the eastern coastline of South America and the western coastline of the African continent. Various hypotheses were proposed, but for over a hundred years no hard data were developed to lead to an acceptance of any of the reasons why the South American coastline seemed to fit the curvatures of the African coastline.

In the 1950s, however, data began to be uncovered implying that the two continents had, in fact, originally been joined together into one colossal land mass. Studies of rock magnetism of the two continents, earthquake epicenters, and stratefication similarities led to the inevitable conclusion that a huge mass of the colossal continent had broken loose and had "drifted" or had been pushed apart by suboceanic forces characteristic of the subterranean zones.

By the mid sixties, the scientific community had charted various fractures and drifts as they had presumably taken place during the past 200 million years. Not only had the South American land mass been joined to the African continent, and then separated and drifted apart, but also other large areas had fractured and were now separated by gulfs and oceans. Furthermore, a most interesting conclusion was that they are continuing to drift -- huge solid continents floating and drifting upon the molten rock underlayers of the earth's crust just as sheets of ice float and drift upon the molten ice we call water.

As one now reads Auden's poem, we see a romantically simplified view of the world in which all is well. As Professor Mason Ellsworth writes, "We see lovers who defy the exigencies of time, singing in a setting that suggests fertility and fullness. The lovers' joy is expressed in a pleasantly foolish rhapsody that sees the world going on forever in happiness entirely within their control."[14]

Another view is equally possible. "The lovers' vow is magnificent, but the railway arch may be intended to signify a comparatively sordid rendevous."[15] If we keep in mind, therefore, this possible scenario, we have cause to wonder about the implications of the lines: "I'll love you, dear, I'll love you / Till China and Africa meet." Because we know of continental drift, are we to infer that the speaker's love is endless because China and Africa will never meet? Or, are we to infer that all is not within the lovers' control? If gigantic homogeneous continents can separate and drift apart, is there anything that is certain and permanent? From an earth science perspective, is there evidence that a river has jumped "over the mountain?" Has an ocean been "hung up to dry?" Is he saying "I want you to believe it is forever, but no one truly knows about tomorrow."

> As I walked out one evening,
> Walking down Bristol Street,
> The crowds upon the pavement
> Were fields of harvest wheat.
>
> And down by the brimming river
> I heard a lover sing
> Under an arch of the railway:
> "Love has no ending.
>
> I'll love you, dear, I'll love you
> Till China and Africa meet,
> And the river jumps over the mountain
> And the salmon sing in the street.
>
> I'll love you till the ocean
> Is folded up and hung to dry,
> And the seven stars go squawking
> Like geese about the sky.

The years shall run like rabbits,
For in my arms I hold
The Flower of the Ages,
And the first love of the world."

But all the clocks in the city
Began to whirr and chime:
"O let not Time deceive you,
You cannot conquer Time.

In the burrows of the Nightmare
Where Justice naked is,
Time watches from the shadow
And coughs when you would kiss.

In headaches and in worry
Vaguely life leaks away,
And Time will have his fancy
Tomorrow or today.

Into many a green valley
Drifts the appalling snow;
Time breaks the threaded dances
And the diver's brilliant bow.

O plunge your hands in water,
Plunge them in up to the wrist;
Stare, stare in the basin
And wonder what you've missed.

The glacier knocks in the cupboard,
The desert sighs in the bed,
And the crack in the tea-cup opens
A lane to the land of the dead.

Where the beggars raffle the banknotes

And the Giant is enchanting to Jack,
And the Lily-white Boy is a Roarer,
And Jill goes down on her back.

O look, look in the mirror,
O look in your distress;
Life remains a blessing
Although you cannot bless.

O stand, stand at the window
As the tears scald and start;
You shall love your crooked neighbor
With your crooked heart."

It was late, late in the evening,
The lovers they were gone;
The clocks had ceased their chiming,
And the deep river ran on.
- "As I Walked Out One Evening" -
- W. H. Auden -

Love, hate, war and pain have been found to have analogies in the earth's phenomena. Still, one of the most interesting analogies is found in the book-length poem entitled *Orchestra*. Written by Sir John Davies (1569-1626), the entire poem is dedicated to the antiquity and universality of dance. He saw dance and music as being the dominent pattern around which the universe, the earth, and all of the constituents were ordered.

One portion of the poem pictures the dance of the oceans. He describes the tides and wave actions as unique dances upon the "broad breast" of the earth. One can observe, too, how he dealt with the idea that "Only the earth doth stand forever still." Carefully he explained that it also moves, and he implied the earth's dance as it "swiftly turneth underneath their feet."

For lo, the sea, that fleets about the land
And like a girdle clips her solid waist,
Music and measure both doth understand;
For his great crystal eye is always cast
Up to the moon and on her fixed fast;
And as she danceth in her pallid sphere,
So danceth he about the centre here.

50
Sometimes his proud green waves, in order set,
One after other flow unto the shore;
Which when they have with many kisses wet,
They ebb away in order as before;
And to make known his courtly love the more,
He oft doth lay aside his three-fork'd mace
And with his arms the timorous earth embrace.

51
Only the earth doth stand forever still:
Her rocks remove not, nor her mountains meet;
(Although some wits enrich'd with learning's skill
Say heaven stands firm and that the earth doth fleet
And swiftly turneth underneath their feet)
Yet, though the earth is ever steadfast seen,
On her broad breast hath dancing ever been.
- from *Orchestra* -
- Sir John Davies -

But if one is to insist that the earth itself stands forever still, such is not the case for its constituents. Not only do the oceans dance, but also the rivers and streams "Observe a dance in their wild wanderings."

52
For those blue veins that through her body spread,
Those sapphire streams which from great hills do spring,

(The earth's great dugs, for every wight is fed
With sweet fresh moisture from them issuing)
Observe a dance in their wild wandering;
And still their dance begets a murmur sweet,
And still the murmur with the dance doth meet.

53
Of all their ways I love Meander's path,
Which to the tunes of dying swans doth dance.
Such winding sleights, such turns and tricks he hath,
Such creeks, such wrenches, and such dalliance,
That, whether it be hap or heedless chance,
In his indented course and wriggling play
He seems to dance a perfect cunning hay.

54
But wherefore do these streams forever run?
To keep themselves forever sweet and clear;
For let their everlasting course be done,
They straight corrupt and foul with mud appear.
O ye sweet nymphs, that beauty's loss do fear,
Contemn the drugs that physic doth devise
And learn of Love this dainty exercise.

- from *Orchestra* -
- Sir John Davies -

Whether or not the rivers were dances upon the earth was not the most serious concern among the earth scientists. Much more perplexing to them were the questions: What was the origin of the seas and rivers? Why was one salty and the other sweet? One vague answer to the question was found in Ecclesiastes: "All the rivers run into the sea; yet the sea is not full; unto the place from whence the river comes, thither shall they return again." But how? Another proposal was reasoned by Plato who taught that deep within the earth were huge caverns filled with perpetually moving water which was the source of all rivers, and it was the reservoir into which all rivers ultimately flowed. But how?

In the first century before the common era, the Roman poet, Titus Lucretius Carus, explained the phenomena using his own logical intuition and analogies to common experience. In earlier lines of *De Rerum Natura*, he had described the earth as originally having a smooth homogeneous surface. It was drastically changed when great chunks were ripped away as the sun and moon were created. The following lines begin by describing how the newly created hollow was filled with the salty water -- "the salt sweat" of the earth. This was followed by a layering of materials of varying densities -- sludge on the bottom, the seas atop the sludge, air atop the seas, and the fiery hotness of the ether atop it all.

>   When sun and moon
> Departed, earth sank suddenly, where now
> The blue-green reach of sea extends; earth's hollows
> Filled with salt water, and day after day
> The more the tidal force of air, the rays
> Of sun beat down on earth with frequent blows,
> Compressing it from width to narrowness,
> Its nature all compact, so much the more
> By the salt sweat exuded from its body
> Its ooze increased the sea, the swimming plains;
> And, likewise all the more, uncounted motes
> Of heat and air escaped, rose far from earth,
> Crowded the shining reaches of the sky,
> Valleys and plains subsided, mountains loomed
> To lofty heights, for the crags could not sink down
> Nor all the parts descend to equal depth.
>
> So, then, the weight of earth solidified,
> And all the heavy sludge of all the world
> Settled like lees or wine-dregs in a cask.
> Then sea, then air, then all the fiery realm
> Of ether, with translucent particles,

> Was purified, lightness varying
> From lightness, till the most delicate
> Floated, most frail, above all whirlwind force,
> Of the least air, above all whirlwind force,
> Unmingling with all turbulence of storm,
> Smoothly the calm direction of its fire
> Gliding along, like to the Pontic sea
> Keeping the even tenor of its way.
> > - from *De Rerum Natura* V, 481-532 -
> > - Lucretius -

During the sixteenth and seventeenth centuries, explanations about the seas, rivers, and rain were tied to the writings of Plato and the scriptures. Scholars continued to explain geological processes as variations of biological processes, just as we have seen biological processes explained as variations of geological processes. In the lines from Sir John Denham's (1615-1669) poem, notice how the explanations are interwoven.

> Our God, when Heaven and earth He did create,
> Form'd Man, who should of both participate;
> If our lives motions theirs must imitate,
> Our knowledge, like our blood, should circulate.
>
> Into earth's spungy veins the ocean sinks,
> These rivers, to replenish which he drinks;
> So learning, which from reason's fountain springs,
> Back to the source some secret channel brings.
> > - from "The Progress of Love" -
> > - Sir John Denham -

Finally, in the late seventeenth century, Pierre Perrault used physical measurements to prove the relationship among the seas, rivers, and rain. Water evaporates from the seas leaving its minute traces of salt; the water vapor moves ashore as clouds; the droplets coalesce and fall as rain; the rains wash across the earth dissolving traces of minerals and filling the rivers; the rivers flow into the seas where again the waters evaporate leaving their salts.

Still, there were those among the poets who were saddened by the realization of such a mechanical, scientific process. They much perferred the more romanticized process as was described by Lucretius.

> Now I'll explain how, in the lofty clouds,
> The moisture gathers and falls to the earth as rain.
> First let me say that out of everything
> Numerous motes of water rise to the clouds
> And there increase, as do the clouds themselves
> With all their contents, growing as our bodies
> Grow with their blood and plasma, sweat and lymph.
> Clouds also soak up liquid from the deep
> Like hanging woolly fleeces borne by the wind
> High over stream or ocean. When the clouds
> Are nearly running over, full to the brim
> With water-seeds, they labor to discharge
> Their moisture in two ways: the force of wind
> May push them in such masses, close together,
> That their compression and compactness cause
> The rain to fall, from the sheer weight of cloud,
> From dense adhesion. But then, also, winds
> May thin the clouds, dissolve them, as it were,
> And the sun's heat precipitate the rain
> As fire melts wax. Rains are most violent
> Under the double pressure of their weight
> And wind's commotion; for the rain pours down
> In more continual fall when cloud on cloud
> Piles high, and mist is borne along with mist,
> And many, many motes and water-seeds
> Are everywhere-even the steaming earth
> Discharging wetness. Finally the sun
> Through the storm-darknesss sends the rays of light

Shining against the cloud-spray, and we see
The glory of the rainbow.
- from *De Rerum Natura* VI, 496-525 -
- Lucretius -

As a final look at how the poet has creatively utilized the earth and its phenomena, we turn to a word-picture of the twentieth century.

its raining womens voices as if they were dead even to memory

its you too that its raining marvelous encounters of my life o droplets

and these rearing clouds go whinneying by a whole universe of auricular towns

listen if its raining while regret and disdain weep an ancient music

hear the bonds fall which hold you above and below

-"It's Raining"-
-Guillaume Apollinaire-

CHAPTER VIII   MATTER

> Clay is moulded to make a vesel, but the utility
> of the vessel lies in the space where there is
> nothing....Thus, in taking advantage of what is,
> we recognize the utility of what is not.
> -from *Tao te ching*-
> -Lao Tse (604-531 B. C.)-

Early man observed that matter could not be destroyed. When he burned wood, the matter vanished save for a small portion of ashes. When iron rusted, it became progressively smaller. When a metal was immersed in acid, it disappeared entirely. Man could destroy matter, and with that as the case, man should also be able to create matter.

Such a conclusion, however, ran contrary to the prevailing religious belief. God created all things. If man could create matter, was man equal to God? If man could destroy matter, which was   God's creation, was man superior to God? The ancient philosophers, therefore, were forced to the conclusion that matter could be neither created nor destroyed. But matter could be changed. Euripedes (484-407 B.C.) expressed the concept in one of his poems. He drew upon observations of nature to prove the everlastingness of matter, and, by inference, to the everlastingness of his soul. We note that he used the word "change," and states clearly that "nothing dies."

> O potent Earth, and Heaven god-built,--
> Of Heaven are god and man begot,
> And Earth brings forth to mortal lot
> Fruits of the rain from Heaven spilt--
>
> The grass she bears and wild things' breed:
> All-Mother rightly is her name.
> To Earth go back from Earth who came,
> And what was born from skyey seed
>
> Travels again to Heaven's field.
> There's nothing dies of all that's born;
> But one by other toss'd and torn
> Old things are changed, and new revealed.
>   - from "Earth and Sky" -
>   - Euripedes -

Lucretius, the Roman poet, also wrote of the concept of everlastingness and the concept of balanced change. His conception was clearly that "The parts must add \ Up to the sum." The world and all of its constituents was finite and, although changing, "never are lost."

>   Do not suppose
> I take too much for granted when I claim
> That earth and fire are mortal, and that air
> And water perish and are born again.
> Take, to begin with, any part of the earth,
> Burnt by continual suns, trampled by hosts,
> Exhaling mist or dust, and flying clouds
> Dispersed by the great gales across the sky.

Part of the soil the heavy rainstorms call
To dissolution; riverbanks are shorn,
Gnawed by the currents. For every benefit
Requital must be given. Earth's our mother,
Also our common grave. And so you see
Earth is receiving loss and gain forever.
No need for words to prove that ocean, streams,
And springs brim over always with new floods;
On every side downrushing mighty waters
Proclaim the fact, but from the surfaces
There's always a subtraction, so excess
Is nullified, wind and sun exert their powers
Of diminution, there's a seepage down
Into the earth, the salt is filtered out,
The substance of the water oozes back
To a confluence at the fountainheads
Of all the rivers, and then flows again
Down the fresh channels of its earlier days.
And air is changed completely, hour by hour,
Moment by moment, in more ways than men
Can ever count; whatever streams from things
Is always poured into that mighty sea,
The ocean of the air; but in its turn
This reassigns to things their particles
Renews them as they flow away. All things
Would otherwise be dissloved and changed to air.
There is no end to the continual process,
Air is always rising out of things, and falling
Back into things again. In other words,
Recurring influence and effluence.
Likewise the generous giver of clear light,
The ethereal sun, forever flooding heaven
With new illumination, light on light,
Changed every single moment--brightness falls
From air, is lost, renewed. This you can tell
By the way clouds begin to veil the sun,

To break, as it were, the rays of light; at once
The lower part of these is gone, the earth
Is dark with shadow where the clouds ride over.
And so, you realize, things forever need
Renewal of shining, every flash of light
Loses intensity, there would be nothing
Visible in the day, did not the sun
Forever stream replenishment. Look about you!
By night we have in our own halls on earth
Our suns, such as they are, the hanging lamps,
The torches flaring bright, or thick with smoke.
What eager servants, always on the go
To keep the light renewed, the wavering flame
Uninterrupted (or at least to seem so),
With spark so quick to follow the extinction
Of prior spark! So sun and moon and stars
We know form the processional of light
In infinite new succession, a mote, a flash,
Another and another, on and on.
But don't believe none of them ever falters.
You see that stones are worn away by time,
Rocks rot, and towers topple, even the shrines
And images of gods grow very tired,
Develop cracks or wrinkles, their holy wills
Unable to extend their fated term,
To litigate against the Laws of Nature.
And don't we see the monuments of men
Collapse, as if to ask us, "Are not we
As frail as those whom we commemorate?"
Boulders come plunging down from mountain heights,
Poor weaklings, with no power to resist
The thrust that says to them, Your time has come!
But they would be rooted in steadfastness
Had they endured from time beyond all time,
As far back as infinity. Look about you!
Whatever it is that holds in its embrace

All earth, if it projects, as some men say,
All things out of itself, and takes them back
When they have perished, must itself consist
Of mortal elements. The parts must add
Up to the sum. Whatever gives away
Must lose in the procedure, and gain again
Whenever it takes back.
>-from *De Rerum Natura* V, 247-327-
>-Lucretius-

For over sixteen centuries this fundamental but scientifically unverified principle that matter could neither be created nor destroyed, continued to form the basis of scientific thought. It left unanswered the question, however, of whether the principle reinforced the credibility of God's Creation, or whether God's Creation dictated the acceptance of the principle. George Wither's belief was typical of many of the seventeenth century poets. His poem, written about 1635 with the unusually long title of "As soone, as wee to bee, begunne; / We did beginne, to be Vndone," described the earth's regenerative cycle and the accepted idea of change. It also implied a parallel to mankind.

When some, of former Ages, had a meaning
An Emblem of Mortality to make,
They form'd an Infant on a Death's-head leaning,
And, round about, encircled with a Snake.
The Child so pictur'd was to signify
That from our very Birth our Dying springs;
The Snake, her Tail devouring, doth imply
The Revolution of all Earthy things.
For whatsoever hath beginning, here
Begins, immediately, to vary from
The same it was; and doth at last appear
What very few did think it should become.
The solid Stone doth moulder into Earth;
That Earth, ere long, to Water rarifies;

> That Water gives an Airy Vapor birth;
> And thence a Fiery-Comet doth arise:
> That moves until itself it so impair
> That, from a burning-Meteor, back again
> It sinketh down, and thickens into Air;
> That Air becomes a Cloud; then Drops of Rain;
> Those Drops, descending on a Rocky-Ground,
> There settle into Earth, which more and more
> Doth harden still; so, running out the round,
> It grows to be the Stone it was before.
> Thus, All things wheel about; and, each Beginning,
> Made entrance to its own Destruction hath.
> The Life of Nature ent'reth in with Sinning;
> And is for ever waited on by Death:
> The Life of Grace is form'd by Death to Sin;
> And there doth Life-eternal straight begin.
>   - "As Soone, as wee to bee, begunne;" -
>   - George Wither -

Still, during the seventeenth century there was an uncomfortableness among the men of science and the theologians. The scientists, or natural philosophers, were uncomfortable with the limits of their knowledge. Thus they continually questioned the why and how of natural phenomena. The theologians, on the other hand, were uncomfortable with the questioning and challenging of accepted dogma by the scientists. The why and how of natural phenomena were the Lord's design; one should accept phenomena as they are found. And on both sides we find the poets.

One of the last writings of the blind poet, John Milton (1608-1674), was the great epic *Paradise Lost*. In this poem we find the clever method he chose to express his own view and to support the theologians' view. The poet does not himself tell man to abandon his concern for the how and why. Instead he simply "reports" the dialog between Adam, the first man, and the angel, Raphael. It is the angel who instructs man through his discourse with Adam. In

Book VIII of *Paradise Lost* Adam questions Raphael about why the universe is as it is.

> .... something yet of doubt remains,
> Which only thy solution can resolve.
> When I behold this goodly frame, this world
> Of heav'n and earth consisting, and compute
> Their magnitudes, this earth a spot, a grain,
> An atom, with the firmament compared
> And all her numbered stars, that seem to roll
> Spaces incomprehensible (for such
> Their distance argues and their swift return
> Diurnal) merely to officiate light
> Round this opacous earth, this punctual spot,
> One day and night; in all their vast survey
> Useless besides; reasoning I oft admire,
> How nature wise and frugal could commit
> Such disproportions, with superfluous hand
> So many nobler bodies to create
> Greater so manifold, to this one use,
> For aught appears, and on their orbs impose
> Such restless revolution day by day
> Repeated, while the sedentary earth,
> That better might with far less compass move,
> Served by more noble than herself, attains
> Her end without least motion, and receives,
> As tribute such a sumless journey brought
> Of incorporeal speed, her warmth and light;
> Speed, to describe whose swiftness number fails."

The angel, Raphael, replies to Adam's questions and in so doing convinces Adam, and quite possibly the reader, that "daily life is the prime wisdom."

> "To ask or search I blame thee not, for heav'n
> Is as the book of God before thee set,

Wherein to read his wondrous works, and learn
His seasons, hours, or days, or months, or years:
This to attain, whether heav'n move or earth,
Imports not, if thou reckon right; the rest
From man or angel the great Architect
Did wisely to conceal, and not divulge
His secrets to be scanned by them who ought
Rather admire; or if they list to try
Conjecture, he his fabric of the heav'ns
Hath left to their disputes, perhaps to move
His laughter at their quaint opinions wide
But whether thus these things, or whether not,
Whether the sun predominant in heav'n
Rise on the earth, or earth rise on the sun,
He from the east his flaming road begin,
Or she from west her silent course advance
With inoffensive pace that spinning sleeps
On her soft axle, while she paces ev'n,
And bears thee soft with the smooth air along,
Solicit not thy thoughts with matters hid,
Leave them to God above, him serve and fear;
Of other creatures, as him pleases best,
Wherever placed, let him dispose: joy thou
In what he gives to thee, this Paradise
And thy fair Eve; heav'n is for thee too high
To know what passes there; be lowly wise:
Think only what concerns thee and thy being;
Dream not of other worlds, what creatures there
Live, in what state, condition, or degree,
Contented that thus far hath been revealed
Not of earth only but of highest heav'n."
To whom thus Adam cleared of doubt, replied.
"How fully hast thou satisfied me, pure
Intelligence of heav'n, angel serene,
And freed from intricacies, taught to live
The easiest way, nor with perplexing thoughts

To interrupt the sweet of life, from which
God hath bid dwell far off all anxious cares,
And not molest us, unless we ourselves
Seek them with wand'ring thoughts, and notions vain.
But apt the mind of fancy is to rove
Unchecked, and of her roving is no end;
Till warned, or by experience taught, she learn,
That not to know at large of things remote
From use, obscure and subtle, but to know
That which before us lies in daily life,
Is the prime wisdom; what is more, is fume,
Or emptiness, or fond impertinence,
And renders us in things that most concern
Unpracticed, unprepared, and still to seek.
      -from *Paradise Lost* VIII, 113-194-
      -John Milton-

    In spite of the angel's admonition, however, the natural philosophers continued to ask how and why. Even during the time of Milton's writing, Jan Baptist van Helmont (1580-1644), a Belgian physician, was attacking the question of the creation of matter. He planted a tree in a container with a weighed amount of soil. Patiently he kept records of the water used and the plant's growth. After a period of some five years, he was able to show that the weight of the tree increased by 164 pounds while the weight of the soil remained within an ounce or so of its original weight. Soil, therefore, was not a significant component of the tree's composition. But rather than to declare that matter had been created, he showed that the tree was composed of carbon dioxide gas from the air and water from the soil. The two substances, a colorless gas and a colorless liquid, were combined and changed to produce a new substance -- solid, woody cellulose. The total weight of the reactants equaled the total weight of the product. Matter was not created.

    But progress toward a definite proof of the indestructability of matter was slow. Another hundred years were to pass until the advent of precise weighing instruments and carefully controlled experimentation provided scientific data to verify that matter was neither created nor destroyed.

If matter were not destroyed, it apparently was degraded into simpler components. The loss of matter, when wood burned or when iron rusted away, must have a commonality. The commonality was defined by the Prussian chemist and physician, George Stahl (1660-1734). He called the substance phlogiston. He taught that phlogiston escaped from matter when it burned, or as with iron, when it rusted. The loss of phlogiston accounted for the loss of matter, and the search for this elusive element began.

But a rude discovery lay ahead. The belief that rust was the product when phlogiston left iron was severely challenged when experimenters found that rust weighed more than the original iron!

Antoine Lavoisier (1734-1794) placed metallic tin and oxygen gas in a sealed glass container, and by heating it changed the shiny metal to powdery white tin oxide. The total weight of the container and its contents remained the same proving that matter was neither created nor destroyed, although the original substances were both chemically and physically changed. His work gave rise to the classic experiment with mercuric oxide, the rust of mercury. The powdered red mercuric oxide was placed in a sealed glass container, was weighed and heated. As the contents became hot, the oxide released its oxygen and became shiny, silver mercury metal. When the heat was removed, the oxygen recombined with the silvery liquid mercury as it cooled and reformed the red powdery mercuric oxide. This reversible reaction, of oxide to metal to oxide, could be repeated over and over without any loss or gain of weight. Undeniably, matter was neither created nor destroyed although it was quite obviously changed.

Walt Whitman could now legitimately exhort in "The Song of Myself" that "My tongue, every atom of my blood, formed from this soil, this air." And Shelley could echo the new truths from science:

>   There's not one atom of yon earth
>   But once was living man;
>   Nor the minutest drop of rain,
>   That hangeth in its thinest cloud,
>   But that flowed in human veins.
>   - from "Queen Mab" -
>   - Percy Bysshe Shelley -

Thus the relationship among creation, destruction, and change of matter was formalized in the eighteenth century to give us what is now known as the Law of Conservation of Mass -- matter is neither created nor destroyed, but its form can be changed. The nineteenth century birthed a similar conclusion regarding energy which was expressed as the Law of Conservation of Energy -- energy is neither created nor destroyed, but its form can be changed. And the twentieth century, in a blinding flash of white light and a thunderous roar at Los Alamos, New Mexico, ushered in the atomic era which proclaimed dramatically that matter can become energy, and energy can become matter, -- but the total will always be the same.

The Chain of Being

A powerful influence consistently molded the pattern of scientific, religious, philosophical, and political thought especially during the seventeenth and eighteenth centuries in Europe. The Chain of Being was that conviction.

The Chain of Being, when expressed most succinctly, described the tenet that all phenomena, all existence, and all material things were interlinked and interdependent. It was the ultimate demonstration and proof of the magnificence of God's Creation.

Quite probably the Chain of Being had its roots in Plato's concept of plentitude -- that all possible kinds of things exist --, and from the Biblical scriptures -- that in the beginning God created everything. Nothing new can appear; nothing in existence can disappear. Reenforcing the concept of plentitude and the scriptures was the idea of the everlastingness of matter -- that it could be neither created nor destroyed. Together, they produced an unshakable belief that nature, from unorganized non-living matter to the most highly organized life, formed an unbroken and metaphysical series expressed as: That which is created never ceases to exist. God's Creation was perfect; why should any part ever change? The Chain formed the perfect circle -- each link joined to the next -- without end.

In Alexander Pope's much quoted "Essay on Man," we see that he goes a step further in describing the Chain of Being as God's Creation, "which from God began." Its mechanism is so precise and its design is so balanced that "one step broken, the great scale's destroyed." "Vile worm!" is his reaction toward anyone who should challenge this perfection.

> Vast chain of Being! which from God began;
> Natures ethereal, human, angel, man,
> Beast, bird, fish, insect, who no eye can see,
> No glass can reach; from infinite to thee;
> From thee to nothing.-- On superior powers
> Were we to press, inferior might on ours;
> Or in the full creation leave a void,
> Where, one step broken, the great scale's
>   destroy'd:
> From Nature's chain whatever link you like,
> Tenth, or ten thousandth, breaks the chain alike.
> And if each system in gradation roll,
> Alike essential to th' amazing Whole,
> The least confusion but in one, not all
> The system only, but the Whole must fall.
> Let earth unbalanced from her orbit fly,
> Planets and stars run lawless thro' the sky;
> Let ruling angels from their spheres be
>   hurl'd,
> Being on being wreck'd, and world on world;

> Heav'n's whole foundations to their centre nod,
> And Nature tremble to the throne of God!
> All this dread order break -- for whom?
> > for thee?
> Vile worm! -- O madness! pride! impiety!
> > -from "Essay on Man" IX, 237-258-
> > -Alexander Pope-

Pope proceeded to caution one against questioning the perfection of Creation as it manifest in the Chain of Being. In spite of inconsistencies and apparent contradictions, "One truth is clear" he wrote, "Whatever is, is right."

> All Nature is but Art unknown to thee;
> All Chance, Direction, which thou canst
> > not see;
> All Discord, Harmony not understood;
> All partial Evil, universal Good:
> And spite of Pride, in erring Reason's spite,
> One truth is clear, Whatever is, is right.
> > - from "Essay on Man" IX, 289-294 -
> > - Alexander Pope -

Pope's message, like that of John Milton, was that one should passively accept the world as it is. His message did not prevail. The scientists and scholars studied the earthly and heavenly phenomena and searched for explanations and reasons for inconsistencies. The unifying belief was strong among scientists that through knowledge man could control his existence, his environment, and his destiny. The result was the development of experimental science which, although it caused a departure from the mystical Chain of Being, brought forth stronger and verifiable evidence of the interrelatedness of phenomena and the dynamics of an everchanging yet constant existence.

In the Far East, however, the relationship between the poet and the scholar had been quite the reverse of the European relationship. The Chinese poet, Hsun Tzu (298-238 B.C.), wrote one of the strongest and most definite challenges to the scholars. He rebuked them for neglecting the potential of human effort, and for unquestioningly admiring and accepting nature. This, he said, was in reality missing "the nature of things." When certain natural factors were seemingly beyond the control of man, it was up to man to adapt himself and to exert his own control upon nature. Yet, natural science did not develop in China. One explanation is that "although Hsun Tzu enjoyed great prestige in the Han dynasty, his theory of overcoming nature was not strong enough to compete with the prevalent doctrine of man and nature, which both Confucianism and Taoism promoted."[1]

> Instead of regarding Heaven as great and admiring it,
> Why not foster it as a thing and regulate it?
> Instead of obeying Heaven and singing praise to it,
> Why not control the Mandate of Heaven and use it?
> Instead of looking on the seasons and waiting for them,
> Why not respond to them and make use of them?
> Instead of letting things multiply by themselves,
> Why not exercise your ability to transform
> [and increase] them?
> Instead of thinking about things as things,
> Why not attend to them so you won't lose them?
> Instead of admiring how things come into being,
> Why not do something to bring them to full development?
>   - Hsun Tzu -

Lest we judge too harshly, however, the inability of the Chinese to construct a system of experimental science, we should perhaps consider a letter written by Albert Einstein (1879-1955) in 1953.

> Development of western science is based upon two great
> achievements; the invention of the formal logical system
> (in Euclidian geometry) by the Greek philosophers, and

the discovery of the possibility to find out causal relationship by systematic experiment (renaissance). In my opinion one has not to be astonished that the Chinese sages have not make these steps. The astonishing thing is that these discoveries were made at all.[2]

The idea of the Chain of Being had a corollary that defined man as a microcosm of the universe. Some of the earliest expressions of the idea can be traced back two thousand years and halfway around the world. The Chinese poet, Tung Chung-shu (179-104 B. C.), wrote: "The ch'i of Heaven is above, the ch'i of earth is below, and the ch'i of man is in between." Professor Chan explained the idea as "Human beings are not only located between Heaven and Earth, but are in correspondence with the natural order. To exemplify this principle, Tung likened the 360 joints in the human body with the 360 degrees by which celestial positions were marked. The flesh and bones of the human body correspond to the texture of the earth. The ears and the eyes correspond to the Sun and Moon, and the five internal organs to the Five Elements. The four limbs were the Four Seasons. Man was a microcosm in which all the contents of the macrocosm were reflected."[3]

To the poets of the sixteenth and early seventeenth centuries, man was as John Donne wrote in "Holy Sonnets", ". . . a little world made cunningly of elements." His contemporary, William Drummond of Hawthornden (1585-1649) elaborated in "Upon the King's Coronation" that "Thou seemst a world in thyself, containing heaven, stars, earth, floods, mountains, forests, and all that lives."

Again, one must continue to remember that God, Creation, the universe, and Man were inseparable. Professor Marjorie Hope Nicolson expressed this well when she wrote, "As (man) grew and flourished, so did his world; as he decayed and died, so did his world. God's pattern was eternally recreated in macrocosm, geocosm, and microcosm. Man's head was a copy of God and the universe, not only in its shape but in its being the seat of Reason. Man, the epitome of God and the world, was rational; so were the world and the universe. Each of the three worlds had its individuality, yet each was involved with the others, and all partook of God."[4] Part of her reasoning, no doubt, was based on poems such as the one written by Thomas Bastard in 1598.

>   Man is a little world and beares the face,
>   And picture of the Vniuersitie (universe):
>   All but resembleth God, all but his glasse,
>   All but the picture of his maiestie.
>   Man is the little world (so we him call)
>   The world the little God, God the great All.
>   - from "De Microcosmo" -
>   - Thomas Bastard -

The early seventeenth century saw poets attempting to draw together more tightly the idea that man was a microcosm of the universe. One must admire the labors of those such as Phineas Flectcher (1582-1650) whose long poem, *The Purple Island*, also known as, "The Island of Man," stretched the parallels and pseudoparallels to their limits. In the first brief passage quoted below, he constructed the analogy that "The whole body (of man) is as it were watered with great plenty of rivers, veins, arteries, and nerves."

>   Canto II, 9
>   Nor is there any part in all this land,
>   But is a little Isle: for thousand brooks
>   In azure channels glide the silver sand;
>   Their serpent windings, and deceiving crooks
>   Circling about, and wat'ring all the plain,
>   Emptie themselves into th' all-drinking main;
>   And creeping forward slide, but never turn again.

Farther along we find him describing the passage of urine

from the kidney to the bladder in much the same way that a geologist might describe the course of a river from its headwaters to the ocean.

Canto II, 24

Down in the vale, where these two parted walls
Differ from each with wide distending space,
Into the lake the Urine-river falls,
Which at the Nephros hill beginnes his race:
    Crooking his banks, he often runs astray,
    Lest his ill streams might backward finde a way:
Thereto, some say, was built a curious frameed bay.

Canto II, 25

The Urine-lake drinking his color'd brook,
By little swells, and fills his stretching sides:
But when the stream the brink 'gins over-look,
A sturdy groom empties the welling tides:
    Spincter some call; who if he loose'd be,
    Or stiffe with cold, out flows the senselesse sea,
And rushing unawares covers the drowne'd lea.

Fletcher's poem proceeds tortuously from system to system, function to function, through what he calls the three provinces -- the head, the heart, and the liver. In Canto IV he expands beyond the analogies of man's parts and the earth's parts to an analogy involving man, earth, and the social creation of man -- the city.

Canto IV, 2

*   *   *   *   *   *   *   *   *

The middle Province next this lower stands,
Where th' Isle's Heart-city spreads its larger
    commands.
Leagu'd to the neighbour towns with sure and
    friendly bands.

Canto IV, 3
Such as that starre, which sets his glorious chair
In midst of heav'n, and to dread darkness here
Gives light and life; such is the citie faire:
Their ends, place, office, state, so nearly neare,
    That those wise ancients from their natures sight,
    And likeness, turn'd their names, and called
        aright
The sunne the great works heart, the heart the lesse
    world's light.
        - from *The Purple Island* -
        - Phineas Fletcher -

John Donne creates a different imagery in his "Elegie XX: Loves Progress." In this poem, "Progress" is understood to mean a journey or expedition, and the poem is understood to mean a journey or expedition, and it is indeed a geographical-sexual expedition in somewhat oblique language. It is sexual in that it is the progression of physical love-making; it is geographical in that it uses analogies of physical geography to the human female body.

Donne's first three lines directly state his basic thesis "that women have only a single proper function, and that the sensible male lover will head directly towards the scene of action, avoiding all diversions from his course. . . . . We are given to believe that nonsensual love is nonsensical."[5]

    Who ever loves, if he do not propose
    The right true end of love, he's one that goes
    To sea for nothing but to make him sick:
        - from "Elegie XX: Loves Progress" -
        - John Donne -

Professor Doniphan Louthan, author and authority on Donne's poetry, describes "Loves Progress" best when she wrote, "The causistical argument deals in statements which would seem highly respectable out of context: 'Perfection is unitie'. If we believe in unity (oneness), we 'prefer one woman'. (Is this) a respectable argument for monogamy? No, we continue the process and prefer one thing in her: her private parts."[6]

> The Nose (like the first Meridian) . . . .
> Sailing toward her India . . . .
> her fair Atlantic Navel stay . . . .
> we love her centrique part.

When the poem is carefully read, we find it to be a craftily constructed merging of earth science with "earthy" science.

> Who ever loves, if he do not propose
> The right true end of love, he's one that goes
> To sea for nothing but to make him sick:
> Love is a bear-whelp born, if we o're lick
> Our love, and force it new strange shapes to take,
> We erre, and a lump a monster make.
> Were not a Calf a monster that were grown
> Face'd like a man, though better then his own?
> Perfection is in unitie: prefer
> One woman first, and then one thing in her.
> I, when I value gold, may think upon
> The ductilness, the application,
> The wholsomness, the ingenuitie,
> From rust, from soil, from fire ever free:
> But if I love it, 'tis because 'tis made
> By our new nature (Use) the soul of trade.
>   All these in women we might think upon
> (If women had them) and yet love but one.
> Can men more injure women then to say
> They love them for that, by which they're not they?
> Makes virtue woman? must I cool my bloud
> Till I both be, and find one wise and good?
> May barren Angels love so. But if we
> Make love to woman; virtue is not she:

As beauty is not nor her wealth: He that strayes thus
From her to hers, is more adulterous,
Then if he took her maid. Search every sphear
And firmament, our Cupid is not there:
He's an infernal god under ground,
With Pluto dwells, where gold and fire abound;
Men to such Gods, their sacrificing Coles
Did not in Altars lay, but pits and holes:
Although we see Celestial bodies move
Above the earth, the earth we Till and love:
So we her ayers contemplate, words and heart,
And virtues; but we love the Centrique part.

 Nor is the soul more worthy, or more fit
For love, then this, as infinit as it.
But in attaining this desired place
How much they erre; that set out at the face!
The hair a Forest is of Ambushes,
Of springes, snares, fetters and manacles:
The brow becalms us when 'tis smooth and plain,
And when 'tis wrinckled, shipwracks us again.
Smooth, 'tis a Paradice, where we would have
Immortal stay, and wrinckled 'tis our grave.
The Nose (like to the first Meridian) runs
Not 'twixt an East an West, but 'twixt two suns;
It leaves a Cheek, a rosie Hemisphere
On either side, and then directs us where
Upon the Islands fortunate we fall,
Not faint Canaries, but Ambrosiall,
Her swelling lips; To which when we are come,
We anchor there, and think our selves at home,
For they seem all: There Syrens songs, and there
Wise Delphick Oracles do fill the ear;
There in a Creek where chosen pearls do swell
The Rhemora her cleaving tongue doth dwell.
These, and (the glorious Promontory) 'her Chin
Ore past; and the streight Hellespont between

The Sestos and Abudos of her breasts,
(Not of two Lovers, but two loves the neasts)
Succeeds a boundless sea, but yet thine eye
Some Island moles may scatter'd there descry;
And Sailing towards her India, in that way
Shall at her fair Atlantick Naval stay;
Though thence the Current be thy Pilot made,
Yet ere thou be where thou wouldst be embay'd,
Thou shalt upon another Forest set,
Where many Shipwrack, and no further get.
When thou art there, consider what this chace
Mispent by thy beginning at the face.
   Rather set out below; practice my Art,
Some Symetry the foot hath with that part
Which thou dost seek, and is thy Map for that
Lovely enough to stop, but not stay at:
Least subject to disguise and change it is;
Men say the Devil never can change his.
It is the Emblem that hath figured
Firmness; 'tis the first part that comes to bed.
Civilitie we see refin'd: the kiss
Which at the face began, transplanted is,
Since to the hand, since to the imperial knee,
Now at the Papal foot delights to be:
If Kings think that the nearer way, and do
Rise from the foot, Lovers may do so too.
For as free Spheres move faster far then can
Birds, whom the air resists, so may that man
Which goes this empty and Aetherial way,
Then if at beauties elements he stay.
Rich Nature hath in women wisely made
Two purses, and their mouths aversely laid:
They then, which to the lower tribute owe,
That way which that Exchequer looks, must go:

> He which doth not, his error is as great,
> As who by Clyster gave the Stomack meat.
> - from "Elegie XX: Loves Progress" -
> - John Donne -

The Little World

Within the context of microcosm, geocosm, and macrocosm, religious-philosophical questions persisted. Did God create a perfect universe because he is perfect? Did God create the perfect universe as the perfect home for his unique creation -- man? Did God even create the universe at all?

First, the question could be asked: why should one expect God's Creation to be perfect? Plato (427-347 B. C.) provided an answer.

> TIMAEUS: Let us therefore state the reason why the framer of this universe of change framed it at all. He was good, and what is good has no particle of envy in it; being therefore without envy he wished all things to be as like himself as possible. This is as valid a principle for the origin of the world of change as we shall discover from the wisdon of men, and we should accept it. God therefore, wishing that all things should be good, and so far as possible nothing be imperfect, and finding the visible universe in a state not of rest but of inharmonious and disorderly motion, reduced it to order from disorder, as he judged that order was in every way better. It is impossible for the best to produce anything but the highest. When he considered, therefore, that in all the realm of visible nature, taking each thing as a whole, nothing without intelligence is to be found that is superior to anything with it, and that intelligence is impossible without soul, in fashioning the universe he implanted reason in soul and soul in body, and so ensured that his work should be by nature highest and best. And so the most likely account

must say that this world came to be in very truth, through god's providence, a living being with soul and intelligence.[7]
- from *Timaeus* I:4 -
- Plato -

We notice that Plato defined the earth as a living-breathing organism, and a thousand years later the idea persisted. To the people of the Elizabethan period of sixteenth century England, "the world was not simply like an animal; it was animate."[8]

One of the first scientific expressions of belief that Earth was alive was by the geologist James Hutton in 1785 in a lecture before the Royal Society of Edinburgh. His contention was that a study of human physiology would provide insight and a better understanding of how and why the Earth functioned as it did. He cited the work of William Harvey, who discovered the circulation system of blood, as providing clues to the cycling of rain, water and the elements. Interestingly, as we shall see later, Harvey derived much of his inspiration from observing the geology of earth -- its oceans, rivers, and tributaries -- microcism and macrocosm, indeed!

Amazingly, two hundred years later, at the dawning of the twenty-first century, the idea still persists among some thoughtful persons that Earth is a complex Being capable of maintaining a habitable environment. This Being has been given the name of the Greek goddess of the earth, Gaia, and although the belief in its existence is not shared widely by geochemists and other scientists, the rationale for its existence has been freely expressed by the proponents.

> The climate and the chemical properties of the Earth now and throughout its history seem always to have been optimal for life. This relative stability persisted despite the sun brightening 30% since Earth formed, volcanoes spewing acid for eons, and green plants polluting the atmosphere with toxic oxygen, among other abuses. For this (a hospitable habitat for life) to

have occured by chance is as unlikely as to survive unscathed a drive blindfolded through rush-hour traffic.[9]

A fundamental principle within the concept of Gaia is the homeostatic balance of the Earth. Relatively constant conditions are seemingly maintained by self-regulating concentrations of inanimate chemicals such as carbon dioxide, nitrogen, and water, and by the complimentary variances of physical conditions, such as temperature, humidity, and radiation.

Carbon dioxide (CO_2) permits solar radiation to penetrate the atmosphere and warm the earth, but CO_2 also inhibits the reflection of the heat back space. It puts a roof over earth much like a greenhouse. Over a period of many years the concentration of CO_2 could increase and the temperature of the earth could grow hotter and hotter. The ice caps would melt and much of the land would be flooded. Ultimately, life would not survive were it not for another characteristic of CO_2.

As the temperature gradually increases, the water in rivers, lakes and oceans evaporates faster, and the concentration of the water molecules in the air (humidity) increases world-wide. Uniquely, CO_2 functions as a nucleus around which water molecules coalesce to form droplets which fall to earth. This rain, acid rain in effect, decreases the concentration of CO_2 in the atmosphere, and this permits more heat to escape from earth into space. A window in the roof of the greenhouse is opened -- the earth cools.

Meanwhile, the CO_2 in the rainwater makes its way into the oceans and is incorporated into the carbonate shells of marine life which ultimately die. Their shells fall to the ocean floor to form compacted layers of sedimentary rock. In thousands of years these rock plates migrate into the zones of volcanic magma and are decomposed. The CO_2 returns to the atmosphere in the gigantic belches of Vesuvius and Mt. St. Helens. The window in the roof of the greenhouse slowly closes. The cycle is complete. It is continuous.

Unquestionably, the consensus is that the habitable

environment of earth results from the various physical, chemical, and biological cycles. But what is the origin of the cycles? Are they cycles of animate Gaia at work? Are they improbable inanimate coincidental phenomena? Or, are they the design of God? For the vast majority of inhabitants of the post Renaissance period in western Europe, their answer was clear -- God!

God is perfect, therefore, His creation is perfect. And it did not go unnoticed to theologians and poets that God's language of the perfect creation was the circle. John Donne wrote in his "Devotions" that "One of the most convenient Hieroglyphicks of God is a circle, and a circle is endless." The circle was the persistent symbol of perfection, and the circles and cycles of nature served as overpowering evidence of God's handiwork. John Milton (1608-1674) symbolically described the Creation:

> He took the golden compasses, prepared
> In God's eternal store, to circumscribe
> The Universe and all created things.
> One foot he centered, and the other turned
> Round through the Vast profundity obscure,
> And said, "Thus far extend, thus far thy bounds;
> This be thy just circumference, O world."
>   - from *Paradise Lost* VII 225-31 -
>   - John Milton -

The circles and cycles within nature were elaborated upon by the poets in their efforts to display the wisdom and godliness behind the Creation. In *De Rerum Natura*, Lucretius described the water cycle as evidence of divine planning. It is interesting to note, however, that no plausible first century explanation was available to explain how the sea water returned to the headwaters of the rivers, or to explain how the salt water of the sea became fresh again upon returning to "its appointed course."

> Men wonder that the sea does not increase
> In size, with so much water pouring down,

> Such confluence of all the streams and rivers;
> Then add the wandering rains, the sweep of storms,
> All sprinkling and all dowsing elements
> That moisten all the lands and seas; include
> Submarine springs - yet all of this adds up
> To hardly more that a drop in that vast deep
> Of ocean's plentitude. No wonder, then,
> The sea does not increase; and anyway
> The sun's evaporating force is strong
> Enough to shrink it somewhat; we have seen
> Clothes, not too well wrung out, hung up to dry
> In the sun's rays. Now there are many seas
> Widespreading, and though the sun from every one
> Takes only a little at a time, it all
> Adds up to a tremendous diminution;
> And winds that sweep across these surfaces
> Can also dry the waters, as we've seen
> Roads harden over night, and mud become
> Dry-caked, with crossing cracks. And I have shown
> That clouds lift water from the ocean's plain
> To sprinkle all the world, now here, now there,
> When the rain falls or winds bring darkening.
> Finally, since a porous quality
> Inheres in earth, wherever it meets the sea,
> Wherever the boundaries join, as water comes
> From earth to sea, as counterflow seeps back
> Out of salt ooze, it loses brackishness,
> Is filtered and distilled, and makes its way
> Once more, all sweet and fresh, in rivulets,
> Channels and streams, to its appointed course.
>   - from *De Rerum Natura* VI, 609-638 -
>   - Lucretius -

For more than a dozen centuries the cycle of water expressed not only the perfection of Creation -- in that nothing is wasted or lost -- but also emphasized the planned orderliness of nature. The following poem, written in 1737, was typical of this belief.

> Rains feed the earth, nor does the earth deny,
> To send it back in vapors to the sky;
> Seas fill the springs, the springs again repay
> Their grateful tribute to the flowing sea;
> Night follows day, seasons the year divide
> 'Twixt winter's nakedness and summers pride.
>   - from "Order: A Poem" -
>   - Anonymous -

Shelly also wrote on the same theme -- the cycle of water. In his poem, "The Cloud", we see how the physical phenomenon of condensed water vapor expressed itself in the first person as though it were a living organism. The cycle per se was emphasized when we realize that the last line, "I arise and build again," returns the reader quite naturally to the beginning -- "I bring fresh showers for the thirsting flowers, / From the seas and the streams." Thus, the poem about the cycle is itself a cycle.

> I bring fresh showers for the thirsting flowers,
>   From the seas and the streams;
> I bear light shade for the leaves when laid
>   In their noonday dreams.
> From my wings are shaken the dews that waken
>   The sweet buds every one,
> When rocked to rest on their mother's breast,
>   As she dances about the sun.
> I wield the flail of the lashing hail,
>   And whiten the green plains under,
> And then again I dissolve it in rain,
>   And laugh as I pass in thunder.
>
> I sift the snow on the mountains below,
>   And their great pines groan aghast;
> And all the night 'tis my pillow white,

While I sleep in the arms of the blast.
Sublime on the towers of my skiey bowers,
    Lightning my pilot sits;
In a cavern under is fettered the thunder,
    It struggles and howls at fits;
Over earth and ocean, with gentle motion,
    This pilot is guiding me,
Lured my love of the genii that move
    In the depths of the purple sea;
Over the rills, and the crags, and the hills,
    Over the lakes and the plains,
Wherever he dream, under mountains or stream,
    The Spirit he loves remains;
And I all the while bask in Heaven's blue smile,
    Whilst he is dissolving in rains.

The sanguine Sunrise, with his meteor eyes,
    And his burning plumes outspread,
Leaps on the back of my sailing rack,
    When the morning star shines dead;
As on the jag of a mountain crag,
    Which an earthquake rocks and swings,
An eagle alit one moment may sit
    In the light of its golden wings.
And when Sunset may breathe, from the lit sea beneath,
    Its ardours of rest and of love,
And the crimson pall of eve may fall
    From the depth of Heaven above,
With wings folded I rest, on mine aery nest,
    As still as a brooding dove.

That orbed maiden with white fire laden,
    Whom mortals call the Moon,
Glides glimmering o'er my fleece-like floor,
    By the midnight breezes strewn;
And wherever the beat of her unseen feet,

  Which only the angels hear,
May have broken the woof of my tent's thin roof,
  The stars peep behind her and peer;
And I laugh to see them whirl and flee,
  Like a swarm of golden bees,
When I widen the rent in the wind-built tent,
  Till the calm rivers, lakes, and seas,
Like strips of the sky fallen through me on high,
  Are each paved with the moon and these.

I bid the Sun's throne with a burning zone,
  And the Moon's with a girdle of pearl;
The volcanoes are dim, and the stars reel and swim,
  When the whirlwinds my banner unfurl.
From cape to cape, with a bridge-like shape,
  Over a torrent sea,
Sunbeam-proof, I hang like a roof,--
  The mountains its columns be.
The triumphal arch through which I march
  With hurricane, fire, and snow,
When the Powers of the air are chained to my chair,
  Is the million-coloured bow;
The sphere-fire above its soft colours wove,
  While the moist Earth was laughing below.

I am the daughter of Earth and Water,
  And the nursling of the Sky;
I pass through the pores of the ocean and shores;
  I change, but I cannot die.
For after the rain when with never a stain
  The pavilion of Heaven is bare,
And the winds and sunbeams with their convex gleams
  Build up the blue dome of air,
I silently laugh at my own cenotaph,
  And out of the caverns of rain,

> Like a child from the womb, like a ghost from the tomb,
>   I arise and unbuild it again.
>   - "The Cloud" -
>   - Percy Bysshe Shelley -

From the smooth melody of the romantic period we come to a harsher, almost abrupt visual poem of the twentieth century. The contemporary poet, Tom Ockerse, treats the water cycle in an entirely different manner of expression. His abstraction requires a second look. The horizon separates the clouds from the water. The C of the word "cloud" visually represents cumulo-nimbus cloud formations, while the letters spill downward as rain. Below the horizon the W of the word "water" creates the reflection of the clouds above, and water is spelled with wavy-lined letters.

But was God's creation perfect? Plato said, yes. The literal theologist said, yes. And poets such as Cowper said, yes.

>   Some say that in the origin of things
> When all creation started into birth,
> The infant elements received a law
> From which they swerve not since. That under force
> Of that controuling ordinance they move,
> And need not his immediate hand, who first
> Prescribed their course, to regulate it now.
> Thus dream they, and contrive to save a God
> The incumbrance of his own concerns, and spare
> The great Artificer of all that moves
> The stress of a continual act, the pain
> Of unremitted vigilance and care,
> As too laborious and severe a task.
> So man the moth is not afraid, it seems,
> To span Omnipotence, and measure might
> That knows no measure, by the scanty rule
> And standard of his own, that is today,
> And is not, ere to-morrow's sun go down.

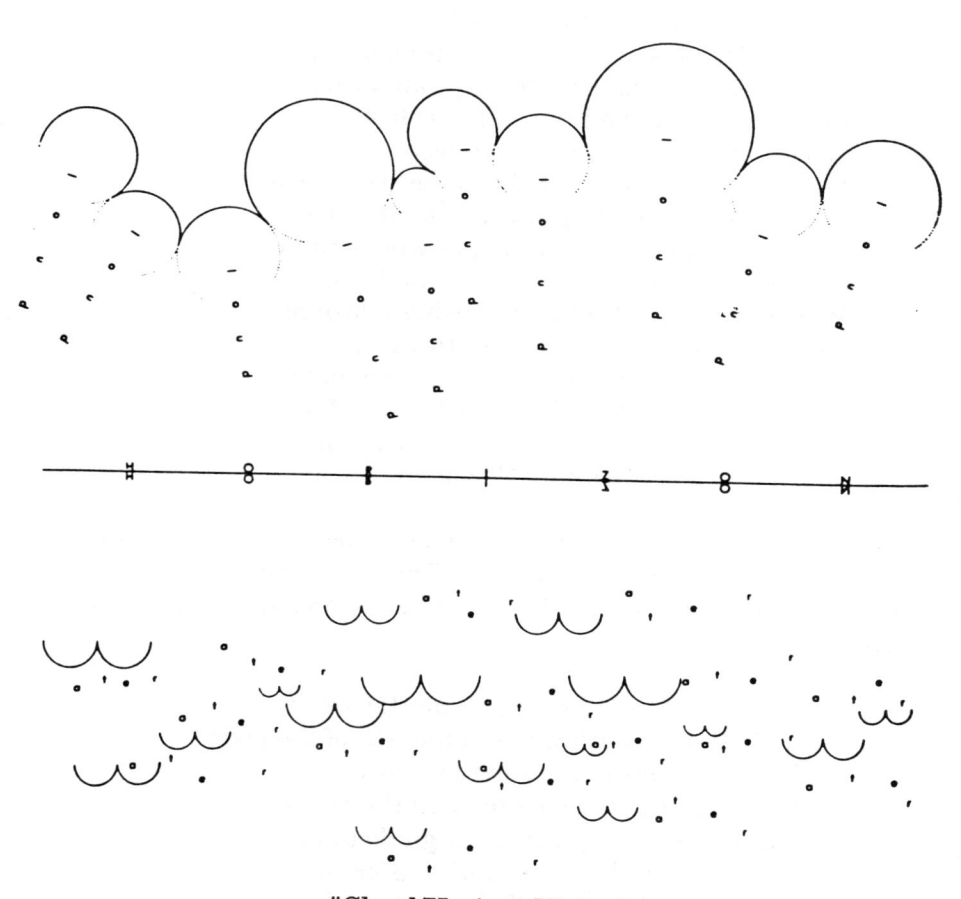

-"Cloud/Horizon/Water"-
-Tom Ockerse-

> But how should matter occupy a charge
> Dull as it is, and satisfy a law
> So vast in its demands, unless impell'd
> To ceaseless service by a ceaseless force,
> And under pressure of some conscious cause?
> The Lord of all, himself through all diffused,
> Sustains and is the life of all that lives.
> Nature is but a name for an effect
> Whose cause is God. He feeds the secret fire
> By which the mighty process is maintain'd,
> Who sleeps not, is not weary; in whose sight
> Slow-circling ages are as transient days;
> Whose work is without labor, whose designs
> No flaw deforms, no difficulty thwarts,
> And whose beneficence no charge exhausts.
> - from "The Task: The Winter Walk at Noon" -
> - William Cowper -

Still others said, no. The poet Lucretius explained the reasons for his rational conclusion. The earth could not have been created by God because, "There are too many things the matter with it."

> Some people do not know how matter works.
> They think that nature needs the will of the gods
> To fit the seasons of the year so nicely
> To human needs, to bring to birth the crops
> And other blessings, which our guide to life,
> The radiance of pleasure, makes us crave
> Through Venus' agency. To be sure, we breed
> To keep the race alive, but to think that gods
> Have organized all things for the sake of men
> Is nothing but a lot of foolishness.

I might not know a thing about the atoms,
But this much I can say, from what I see
Of heaven's ways and many other features:
The nature of the world just could not be
A product of the gods' devising; no,
There are too many things the matter with it.
I'll give you further details, Memmius, later.
             - from *De Rerum Natura* II, 168-184 -
             - Lucretius-

True to his word, Lucretius gives further details. He charges that the erosion of landforms, the unpredictability of weather, the frequency of natural disasters, and diseases are prima facie evidence that God had no part in the earth's creation, or of the managing of it since then!

This world of ours was not prepared for us
By any god. Too much is wrong with it.
For one thing, what the mighty swirl of the sky
Protects, the covetous mountains and the jungles
Have seized a part of, and the cliffs and swamplands
Appropriate their share; and then there's ocean
Keeping the shores as wide apart as may be.
Two-thirds of what there is, that pair of thieves,
Fierce heat, insistent cold, have robbed men of;
And what is left, nature, as violent
As either one, would occupy and homestead
With fence of briar and bramble, but men resist
For their dear lives, groan as they heave the mattock
The way they know they must, or break the soil
Shoving the plow along. Were this not done,
This plowshare-turning of the fertile clod,
This summoning to birth, nothing at all
Could, of its own initiative, leap forth
Into the flow of air. How many a time
The produce of great agonies of toil.

> Burgeons and flourishes, and then the sun
> Is much to hot and burns it to a crisp;
> Or sudden cloudbursts, zero frosts, or winds
> Of hurricane force are, all of them, destroyers.
> And why does nature feed and multiply
> The dreadful race of predatory beasts,
> Man's enemies on sea and land? And why
> Must every season bring disease? And why
> Is early death so free to walk the world?
> When nature, after struggle, tears the child
> Out of its mother's womb to the shores of light,
> He lies there naked, lacking everything,
> Like a sailor driven wave-battered to some coast,
> And the poor little thing fills all the air
> With lamentation -- but that's only right
> In view of all the griefs that lie ahead
> Along his way through life.
> - from *De Rerum Natura* V, 100-228 -
> - Lucretius -

But Lucretius' argument did not prevail. The random collision of "chaotic motes" was not an acceptable explanation of creation for western civilization. Genesis could not be disregarded, and the perfection of the Creation was firmly encapsulated in theological dogma. But if the Creation was perfect, why had earth's countenance changed, asked the geologists? Why have new species appeared and some disappeared, asked the biologists? "Uniformitarianism," answered the geologists; "evolution," answered the biologists. Predictably, an acceptable middle ground between the extreme opinions about the earth's history appeared.

> "When at the word of God the earth emerged, its surface
> was not irregular as we see it now, with mountains, the
> 'warts of earth' as many called them, and profound
> depths, the 'pock-holes.' At the creation the globe of
> earth was a true sphere, unmarked by the blemishes of

height and depth, smooth, equal, even. The gross irregularities of the earth's surface, with its warts and pock-holes, were the abiding evidence of the sin of man. With each of man's sins, the earth had grown increasingly ugly."[10]

The idea of the earth's original perfection which was then followed by its general decay brought about by Adam's sin was acceptable to the theologians. At the same time it removed the religious restriction placed upon the scientists and permitted them to search for the how and why of natural phenomena.

The imperfection of the universe also had a comforting effect. Alexander Pope found it to be a way for man to live with his own imperfections. If God's Creation was obviously not perfect, why should man feel badly about his own lack of perfection?

> Ask for what end the heav'nly bodies
> shine,
> Earth for whose use,-- Pride answers,
> "Tis for mine:
> For me kind Nature wakes her genial
> power,
> Suckles each herb, and spreads out ev'ry
> flower;
> Annual for me the grape, the rose, renew
> The juice nectareous and the balmy dew;
> For me the mine a thousand treasures
> brings;
> Seas roll to waft me, suns to light me rise;
> My footstool earth, my canopy the
> skies.'
> But errs not Nature from this gracious
> end,
> From burning suns when livid deaths de-
> scend,
> When earthquakes swallow, or when tem-

>   pests sweep
> Towns to one grave, whole nations to the
>   deep?
> 'No,' 't is replied, 'the first Almighty
>   Cause
> Acts not by partial but by gen'ral laws;
> Th' exceptions few; some change since all
>   began
> And what created perfect?' -- Why then
>   Man?
> If the great end be human happiness,
> Then Nature deviates; and can Man do
>   less?
>
>   - from "Essay on Man" V, 131-150 -
>   - Alexander Pope -

We have seen how Fletcher and other poets used analogies of man, earth, and the universe to show oneness, and in some way to weld all material being into God's unified Creation. Yet these strained efforts to show congruence were not limited to poets. Some highly respected scientists themselves harbored this belief of oneness; typical of them was Kepler who wrote:

> As the body produces hair on the skin, so the earth produces plants and trees, and as in the former lice are generated, so in the latter caterpillars, crickets, and many other insects and sea-monsters. As the body exudes moisture in tears and sweat through the pores, so does the earth exude amber and bitumen. As urine from the bladder, rivers flow from the mountains. As the body discharges winds that reek of sulphur in subterranean fires, thunder and lightening. As in the veins of living beings are formed blood and sweat, exuded through the passages of the body, so in the veins of the earth are metals and petrifactions, and from them issue steamy torrents. As other living beings take into

their bodies food and drink, the earth through its channels draws into itself stuff of which much is concocted. It swallows the waters of the seas so that the ocean, in spite of the constant flowing of rivers, never overflows.

> - from *Harmonice Mundi* -
> - Johannes Kepler -

The serious belief in the oneness of man and earth frequently served as a deterrent to the progress of science. William Harvey (1578-1657), who described correctly the circulation of the blood, was seriously inhibited as he struggled with the microcosm-macrocosm concept and the Circle of Perfection, not to mention the attacks by the followers of the ancient Greek physician, Galen, whose teachings he challenged. Even in his final writings, Harvey showed clearly the influences that the ancient philosophers had on his thought pattern. He described at one point an intellectual dilemma.

> . . . I tremble lest I have mankind at large for my enemies, so much doth wont and custom, that becomes as another nature, and doctrine once sown and that hath struck deep root, and respect for antiquity influence all men; still the die is cast, and my trust is in my love of truth, and the candour that inheres in cultivated minds.[12]

He went further in expressing his thought processes by referring again to the influences of the prevailing philosophies when he wrote: "I began to think whether there might not be a MOTION, AS IT WERE, IN A CIRCLE. (Harvey's capitalization). Now, this I afterwards found to be true."

> Which motion we may be allowed to call circular, in the same way as Aristotle says that air and the rain emulate the circular motion of the superior bodies; for

the moist earth, warmed by the sun, evaporates; the vapours drawn upwards are condensed, and decending in form of rain, moisten the earth again; and by this arrangement are generations of living things produced....And so, in all likelihood, does it come to pass in the body, through the motion of the blood; the various parts are nourished....when it (blood) returns to the sovereign heart....Here it resumes its due fluidity and recieves an infusuin of natural heat....and is impregnated with spirits....and thence it is again dispersed.[13]

After describing in detail his conclusion regarding the movement of the blood through the arteries and veins, and the dependence of the movement on the action of the heart, he reverted to the ancient philosophy that "The heart, consequently, is the beginning of life; the sun of the microcosm, even as the sun in turn might well be designated as the heart of the world."[14]

The new science of the seventeenth century however, was the tolling of the bell for "man as a little world." The idea did not die quickly, but its erosion by the scientists and their experimental methods was as certain as was the erosion of the white chalk cliffs of Dover by the relentless pounding of the Channel seas.

CHAPTER IX   RELATIVITY AND TRUTH

> Tis only God can know
> Whether the fair idea thou dost show
> Agrees entirely with his own, or no.
>   - from "Leviathon" -
>   - Robert Crowley -

Relativity

In 1905, a previously obscure patent clerk changed man's understanding of the universe. Albert Einstein (1879-1955) presented to the world his Theory of Special Relativity.
   The theory consisted of highly sophisticated mathematical descriptions and proofs of time, space, mass, motion, and gravitation, and described their interrelatedness. Although a precise understanding of the Special Theory was beyond the capacity of most of the public, the theory was generalized in such a way as to popularize the term "relativity," and to incorporate it into both common and uncommon usage to the extent that it impacted not only the sciences but also the arts, poetry, philosophy,

and the social sciences. A poet captured the essence of the popularized relativity in its lowest common denominator -- the truth depends upon one's point of view.

> The world is very flat -
> There is no doubt of that.
> > -"Night Thoughts of a Tortoise
> >   Suffering from Insomnia on a Lawn"-
> - E.V. Rieu -

Einstein's proposal of the Theory of Special Relativity was an apparent disproof of the old laws of absoulte space and the constancy of matter and time. He showed that although Newton's Laws of Motion were correct, they were true only at velocities much less than the speed of light. He showed also that space and time are inseparable; the speed of light is the limit for all natural bodies; and that space, time, and mass are relative to the observer. Mathematically, his equation:

$$E=MC^2$$

stated with utmost clarity that energy, mass, and the velocity of light is the balance beam of the universe. Evergy can be neither created nor destroyed; mass can be neither created nor destroyed; but energy and mass are interconvertible.

Einstein sowed new seeds in the garden of truth, but were the seeds flowers or weeds?

> What was our trust, we trust not,
> > What was our faith, we doubt;
> Whether we must or not
> > We may debate about.
> The soul, perhaps, is a gust of gas
> > And wrong is a form of right --
> But we know that Energy equals Mass
> > By the Square of the Speed of Light.

> What we have known, we know not,
>   What we have proved, abjure.
> Life is a tangled bowknot,
>   But one thing still is sure.
> Come, little lad; come, little lass,
>   Your docile creed recite:
> "We know that Energy equals Mass
>   By the Square of the Speed of Light."
>   - "$E=MC^2$" -
>   - Morris Bishop -

Archibald MacLeish attempted to capture the essence of the man in his 1929 poem, "Einstein". He described a brilliant mind working to penetrate the mysteries of the universe, but he also implied a warm, decidedly human man. The poem communicates a sense of the loneliness and frustration of man's inability to understand his relationship to the universe.

MacLeish's first setting pictures Einstein as a sensitive inheritor of this earth. He is small, determined, and firmly planted in both the abstractions of creativity and the reality of existence.

> Standing between the sun and moon preserves
> A certain secrecy. Or seems to keep
> Something inviolate if only that
> His father was an ape.
>   Sweet music makes
> All of his walls sound hollow and he hears
> Sighs in the paneling and underfoot
> Melancholy voices. So there is a door
> Behind the seamless arras and within
> A living something:-- but no door that will
> Admit the sunlight nor no windows where
> The mirror moon can penetrate his bones
> With cold deflection. He is small and tight
> And solidly contracted into space

>   Opaque and perpendicular which blots
>   Earth with its shadow. And he terminates
>   In shoes which bearing up against the sphere
>   Attract his concentration,
>
>   for he ends
>
>   But it seems assured he ends
>   Precisely at his shoes in proof whereof
>   He can revolve in orbits opposite
>   The orbit of the earth and so refuse
>   All planetary converse. And he wears
>   Cloths that distinguish him from what is not
>   His own circumference, as first a coat
>   Shaped to his back or modeled in reverse
>   Of the surrounding cosmos and below
>   Trousers preserving his detachment from
>   The revolutions of the stars.

MacLeish's second setting describes Einstein's vision of space and its worlds as "rippling ether and swarming motes," and implies the paradox of a human mind in a mindless universe. The scientist's thoughts take him among a multitude of possibilities, "And he moves \ Here as within a garden . . . " Once he discovers truth ". . . the bubble of the world\ Takes center and the circle round his head\ Like golden flies in the summer . . . . " But when his momentary grasp of truth disintegrates he feels "The planet plunge beneath him . . . . "[1] He faces the realization of the paradox of speaking of the truth when the Truth is not known.

>   His hands
>   And face go naked and alone converse
>   With what encloses him, as rough and smooth
>   And sound and silence and the intervals
>   Of rippling ether and the swarming motes
>   Clouding a privy: move to them and make

Shadows that mirror them within his skull
In perpendiculars and curves and planes
And bodiless significances blurred
As figures undersea and images
Patterned from eddies of the air.
      Which are
Perhaps not shadows but the thing itself
And may be understood. *Einstein provisionally*
         before a mirror accepts
         the hypothesis of
         subjective reality

       Decorticate
The petals of the enfolding world and leave
A world in reason which is in himself
And has his own dimensions. Here do trees
Adorn the hillside and hillsides enrich
The hazy marches of the sky and skies
Kindle and char to ashes in the wind,
And winds blow toward him from the verge, and suns
Rise on his dawn and on his dusk go down
And moons prolong his shadow. And he moves
Here as within a garden in a close
And where he moves the bubble of the world
Takes center and there circle round his head
Like golden flies in summer the gold stars.
          ...rejects it

Disintegrates.
      For suddenly he feels
The planet plunge beneath him, and a flare
Falls from the upper darkness to the dark
And awful shadows loom across the sky
That have no life from him and suns go out
And vivid as a drowned man's face the moon
Floats to the lapsing surface of the night

And sinks discolored under.
Less than a world and must communicate
Beyond his knowledge.

In still another scene, the poet seems to be describing Einstein's wrestling with the concept that becomes popularly known as the unified field theory. [2] Signi Lenea Falk interprets this solitary figure of the scientist as the inevitable result of modern physics. The scientist gains control of the universe, but loses its reality in the abstractions. Does reality exist in that which can be described only in terms of mathematical abstraction?[3]

         He lies upon his bed
Exerting on Arcturus and the moon
Forces proportional inversely to
The squares of their remoteness and conceives
The universe.
    Atomic.
        He can count
Ocean in atoms and weigh out the air
In multiples of one and subdivide
Light to its numbers.
        If they will not speak
Let them be silent in their particles.
Let them be dead and he will lie among
Their dust and cipher them -- undo the signs
Of their unreal identities and free
The pure and single factor of all sums --
Solve them to unity.
      - from "Einstein" -
      - Archibald MacLeish -

In one of Einstein's qualitative explanations of Relativity, he used a description of a moving railroad car and a falling object. He explained that if an observer's frame of reference was the inside of the car, the object could be described as falling straight down to the floor. He also explained, however, that if the observer were positioned outside on the station platform as the railroad car sped past, the trajectory would appear as a curve. As the object dropped vertically because of the forces of gravity, it also moved horizontally because of the horizontal movement of the train. Thus two observations of the falling object -- one being a straight line and the other a curved line -- were both correct yet were also contradictory. What then was the truth? Truth depended upon the observer. A revolutionary concept!

Oscar Williams (1900-1964) used a similar setting in his nightmarish fantasy, "The Leg in the Subway." The rider/speaker observes the leg of woman extending beyond one of the coach's interior partitions. All that can be seen is the leg detached from its reality. The leg starts nowhere and goes nowhere. The foot remains motionless. Yet, the subway car moves; the floor is both motionless and moving: " the ankle . . . going nowhere" -- but yet somewhere. What is reality?

The speaker is "wrenched out of the reality of the subway ride." Does society differ from this leg in the subway? When viewed from within, society is going nowhere. Life, relationships, and existence remain constant within their proximities. When viewed from without -- when one is wrenched from the inner realities -- one sees movement, change, and purpose. Just as the passenger appears motionless in the subway car, she exits to a place far distant from where she began. So too does the individual appear motionless in the moving society, unaware of the distance traveled until she exists.

Can truth be seen from the inside, or only from the outside? Or do both truths exist simultaneously?

> When I saw the woman's leg on the floor of the subway train,
> Protrude beyond the panel (while her body overflowed my
>   mind's eye),
> When I saw the pink stocking, black shoe, curve bulging with
>   warmth,

The delicate etching of the hair behind the flesh-colored
    gauze,
When I saw the ankle of Mrs. Nobody going nowhere for a
    nickel,
When I saw this foot motionless on the moving motionless
    floor,
My mind caught on a nail of a distant star, I was wrenched
    out
Of the reality of the subway ride, I hung in a socket of
    distance:...

It spoke saying: To whom does this leg belong? Is it a
    bonus leg
For the rush hour? Is it a forgotten leg? Among the many
Myriads of legs did an extra leg fall in from the Out There?

The planetary approach of the next station exploded atoms of
    light,
And when the train stopped, the leg had grown a surprising
    mate,

I perceived through the hole left by the nail of the star in my
    mind
How civilization was as dark as a wood and dimensional
    with things
And how birds dipped in chromium sang in the crevices
    of our deeds.
      - from "the Leg in the Subway" -
      - Oscar Williams -

  The subjective relativity of distance is played upon by Jeremy Kingston, the contemporary British poet. He describes the awful distances between the bodies within the universe -- incredible distances which defy physical measurement, and certainly our ability to reach out and make contact. And yet, tragically, " . . . they are nearer," says the speaker, " . . .Than I to this my brother,

\ The dark boy...."

> On the hill's top we stood
> Neck back to see which star
> Stood farthest; we had heard
> Some lights made turns so far
> Through such deep space no man
> Knew of them; we had read
> Some lights could not be known.
> "But they are nearer," I said,
> "Than I to this my brother,
> The dark boy of my lies,
> Than him the great stars bend over
> In inches from my eyes."
>
> Indeed, man may not pierce
> Those bending circles, might
> Conceivably not reach
> Even that first dome's light;
> Yet this dark friend, my brother,
> Sets close to me his face,
> And I could force that outer
> Sky before the space
> That bars me from him. Eyes
> Look on me as I start
> My hand to him. He says
> Nothing. I depart.
> - "Distances" -
> - Jeremy Kingston -

A similar contrast was drawn by Robert Frost. Whereas Kingston called attention to the distances between the souls of men, Frost chose to take a more introspective consideration. He seemed to shrug off the emptiness of the vast universe as being personally irrelevant when compared to his own emptiness.

> Snow falling and night falling fast oh fast
> In a field I looked into going past,
> And the ground almost covered smooth in snow,
> But a few weeds and stubble showing last.
> The woods around it have it -- it is theirs.
> All animals are smothered in their lairs.
> I am too absent-spirited to count;
> The loneliness includes me unawares.
> And lonely as it is that loneliness
> Will be more lonely ere it will be less--
> A blanker whiteness of benighted snow
> With no expression, nothing to express.
> They cannot scare me with their empty spaces
> Between stars -- on stars where no human race is.
> I have it in me so much nearer to home
> To scare myself with my own desert places.
>   - "Desert Places" -
>   - Robert Frost -

One might pause to reflect upon the impact of relativity upon others within the scientific community. William Pallister, scientist-poet, reiterated the perspectives of relativity: "The world within, the world without." He calls the reader's attention to one's existence and to being "Dimensionless again!" He develops a series of contrasts that ask the question: "Can all these be the same?"

> I
> There are two worlds beyond a doubt
> Which to all minds appear,
> The world within, the world without,
> The worlds of there and here.
>
> II
> We see the sun rise? see it set?

Illusion lurks in all.
The long set sun is rising yet
And other dawn birds call.
The ancient world, the world of youth
Soon in perspective join.
Our world of doubt, our world of truth,
Are two sides of the coin.

III

The earth has placed on you and me
This four dimensional chain
Of space and time; then we are free,
Dimensionless again!
Here warfare, ignorance and night
Are joined in all but name
With rainbow, sunshine and delight.
Can all these be the same?
A world of earthquakes, floods and storms,
Of sadness and regret,
Of beauty in its million forms
And ultra-violet.

Pallister the poet takes the opportunity to express the views of Pallister the scientist. In section XIV he uses Einstein's classical experiment to build a case for a limited universe. Einstein had set about to prove that light was a particle of matter -- a photon. If it were a particle of matter, then its behavior should correspond to that of any other particle. It should, for example, be attracted to any other body as Newton had so clearly shown. In Einstein's experiment, he predicted that just prior to an eclipse, the light coming from a remote point in space would be bent just as it passed close to the eclipsing body, and such was the case. Pallister uses the idea of the bending path of light to substantiate his thesis of the bounded, limited universe: "A finite world to be defined!"

XIV

Has the great universe an end,
    As there would seem to be?
Light rays in passing our sun seem to bend,
    Respond to gravity.
Men have proved this in an eclipse
    Measured the arc and bend
And, since all arcs form an eclipse
    The whole must have an end.
A finite world to be defined!
    In this we shall not fail;

        - from "Relativity" -
        - William Pallister -

    A careful reading of Pallister's poem calls to mind a certain similarity of thought found in William Blake's poem, "The Tyger." Recall that Pallister asks the question: "Can all these be the same?" Blake also brings together a series of contrasts. As his poem is read, consider his similar and haunting question: "Did he who made the Lamb make thee?"

    Beyond the literal meanings of the words, Blake insinuated into the poem his dislike for the new science as championed by Newton and the new breed of experimental scientists. To Blake, the Tyger symbolized science. It was conceived in the fallen world of reason, truth, and beauty. It was the progeny of the mechanical laws, and, implicit in the descriptive middle verses of the poem, its mere existence caused simultaneous reactions of awe and terror.

    One can sense Blake's anguish as he struggled with his conviction that God created all. How could this be? What chance had reason and beauty (the Lamb) when faced by science (the Tyger)? Blake asks incredulously, "Did he smile his work to see? / Did he who made the Lamb make thee?"

    Tyger! Tyger! burning bright
    In the forests of the night,

What immortal hand or eye
Could frame thy fearful symmetry?

In what distant deeps or skies
Burnt the fire of thine eyes?
On what wings dare he aspire?
What the hand dare seize the fire?

And what shoulder, and what art,
Could twist the sinews of thy heart?
And when thy heart began to beat,
What dread hand, and what dread feet?

What the hammer? what the chain?
In what furnace was thy brain?
What the anvil? what dread grasp
Dare its deadly terrors clasp?

When the stars threw down their spears,
And watered heaven with their tears,
Did he smile his work to see?
Did he who made the Lamb make thee?

Tyger! Tyger! burning bright
In the forests of the night,
What immortal hand or eye,
Dare frame thy fearful symmetry?
- The Tyger -
- William Blake -

The Great War of 1914 to 1918 was for many artists and writers the ultimate evidence of the moral bankruptcy of the intellectual, cultural and social systems of Europe. Religion, rational thought and human values all seemed to be inferentially contradicted by the slaughter of millions of people at the hands of civilized nations.

The alienation among certain writers and artists on the continent grew as popular acceptance of relativity increased, and along with it the realization of its implied elimination of absolute truth. From within the seed-bed of alienation, the counter-culture Dadaists germinated.[4]

The Zurich based movement rejected the standard values of society and set about deliberately to dismantle the arts. The Dadaists created senseless and shocking works as protests against convention. The visual arts were especially vulnerable. Paintings such as a mustachioed Mona Lisa with obscene captions were typical of the zealots.

Although the movement's efforts were ostensibly destructive, they did give birth to the surrealistic school of painting. The Dadaist's positive and lasting contribution to literature, however, has been judged to be much less profound. The piece on the following page is by the German poet, Hugo Ball (1886-1927), who was one of the most active proponents of Dadaism in Europe. Poems of this type "were meant to explode language, and thus bring about a 'Revolution of the Word' similar to the cubist revolution in art."[5]

Whereas many authorities consider the Dadaist movement to have been a dead-end, one cannot overlook the contemporary form of experimental poetry birthed in the 1950s, and referred to as "concrete poetry." Emmett Williams defined it as "poetry is what poets make." His definition is far from a concrete explanation, but one may gain a better understanding of his meaning by returning to his poem "Like Attracts Like" on page 38 in the chapter dealing with the atom.

Mike Weaver, a British art critic, described concrete poetry as "an arrangement of materials according to a scheme or system set up by the poet." One of the practicing concrete poets, Mary Ellen Soet, possibly described it best by saying that "the concrete poet is concerned with making an object to be perceived rather than read." In any case the concrete poets hold the conviction that the old grammatical, syntactical structures are no longer adequate for

KARAWANE

jolifanto bambla ô falli bambla
grossiga m'pfa habla horem
égiga goramen
higo bloiko russula huju
hollaka hollala
anlogo bung
blago bung
blago bung
bosso fataka
ü üü ü
schampa wulla wussa ólobo
hej tatta gôrem
eschige zunbada
wulubu ssubudu uluw ssubudu
tumba ba- umf
kusagauma
ba - umf

      -"KARAWANE"-
      -Hugo Ball-

communicating "advanced processes of thought."[6]

Probably the influence of relativity is not more vividly expressed than in the poem by the contemporary American poet Carl Fernbach-Florsheim. His poem is simply a page torn from his desk calendar. The reader sees the calendar page with "Saturday August 27" across the top. In the center of the page is written in cursive the words "They are setting up new rules - a smaller particle was discovered." It forces the reader to participate within the poet's inner thoughts. "The calendar page is more important," Florsheim says, "than any other piece of paper with the words written or printed on it on the poet's desk on Saturday, August 27; for when the scientist finds it necessary to make a revision of the rules, so does the poet."[7] Did relativity replace truth?

Truth

Poetry written in the 1920s did not attempt directly to exploit the suggestions made by the Theory of Relativity, but a certain amount of experimental writing can be explained by reference to it as has been shown. The real exploitation, however, seems to have developed as popular skepticism about all human knowledge.

Skepticism about knowledge and the role of the scientist began well before Einstein and relativity. Since the seventeenth century much criticism had been leveled against experimental science and its rebuking of Divine Revelation as the source of truth. John Donne (1571-1633) had lamented the conflict, and described the confusion the new science, or "new Philosophy" as he called it, had caused.

> And new Philosophy call all in doubt
> The Element of fire is quite put out;
> The Sun is lost, and th' earth and no man's wit
> Can well direct where to look for it.
>   - from "The First Anniversary" -
>   - John Donne -

Alexander Pope (1688-1744) also showed his concern with the growing conflict over the source of truth. He recognized man's increasing knowledge, but at the same time was increasingly dubious of man's ability to manage his newly found powers over nature and to control his destiny.

> Go, wondrous creature! Mount where Science guides,
> Go, measure earth, weigh air, and state the tides;
> Instruct the planets in what orbs to run,
> Correct old Time, and regulate the Sun;
> Go, teach Eternal Wisdom how to rule -
> Then drop into thyself, and be a fool!
>   - from "Essay on Man" Epistle II -
>   - Alexander Pope -

The newly emerging scientists of the seventeenth century did not escape the sting of the satirical pen of Samuel Butler (1612-1680). The pride of King Charles II, Butler never hesitated to speak out when he saw hypocrisy or ridiculousness. His writings, both verse and prose, were simultaneously characterized by acid and accuracy. His satirical pen found a home when he turned toward the scientists and their self-appointed role as the emissaries of truth.

His poem, "The Elephant in the Moon," was especially popular and focuses upon not only the scientists -- "most profound and wise," but also upon their societies "The glory of a foreign state." Butler believed that vanity and pomposity were the ingredients of their ridiculousness. In all too many cases, their truth was not what was, but what they wished it to be. The poem sarcastically described the telescopic exploration of the Moon by members of a "learn'd society."

> A learn'd Society of late,
> The glory of a foreign state,
> Agreed, upon a summer's night,

To search the Moon by her own light:

The telescope was set in place, whereas the ranking "Virtuoso then in chief" took his position at the tube. He immediately made a series of startling conclusions -- remarkably done without the encumbrance of facts.

> Quoth he: "Th' inhabitants of the Moon
> Who, when the Sun shines hot at noon,
> Do live in cellars underground
> Of eight miles deep and eighty round --
> In which at once they fortify
> Against the Sun, and th' enemy --
> Which they count towns and cities there;
> Because their people's civiler
> Than those rude peasants, that are found
> To live upon the upper ground,
> Called Privolvans, with whom they are
> Perpetually in open war,
> And now both armies, highly 'raged,
> Are in a bloody fight engaged;
> And many fall on both side slain
> As by the glass 'tis clear and plain.

Other "great philosophers" took their turns at the tube and elaborated upon the formers' conclusions, until without warning an astounding discovery was made.

> Quoth he, "A stranger sight appears,
> Than e'er was seen in all the spheres;
> A wonder more unparallel'd,
> Than ever mortal tube beheld;
> An elephant from one of those
> Two mighty armies is broke loose,
> And with the horrour of the fight
> Appears amaz'd, and in a fright:

> Look quickly, lest the sight of us
> Should cause the startled beast t' imboss.
> It is a large one, far more great
> Than e'er was bred in Afric yet,
> From which we boldly may infer,
> The Moon is much the fruitfuller.

They were made jubilant by the discovery "That elephants are in the Moon." This was "...so miraculous a story" that all of their previous stupidities would be forgotten, and their eminence restored.

> This one discovery's enough
> To take all former scandals off --
> But since the world's incredulous
> Of all our scrutinies, and us,
> And with a prejudice prevents
> Our best and worst experiments,
> (As if they were destin'd to miscarry,
> In consort try'd, or solitary)
> And since it is uncertain when
> Such wonders will occur again,
> Let us as cautiously contrive
> To draw an exact narrative
> Of what we every one can swear
> Our eyes themselves have seen appear,
> That, when we publish the account,
> We all may take our oaths upon 't."

During the period of the scientists self-congratulations and mutual admiration, an uneducated lad discovered the truth.

> But while they were diverted all
> With wording the memorial,
> The footboys, for diversion too,
> As have having nothing else to do,

> Seeing the telescope at leisure,
> Turn'd virtuosi for their pleasure;
> Began to gaze upon the Moon,
> As those they waited on had done.
>
> Saw something in the engine creep,
> And, viewing well, discover'd more
> Than all the learn'd had done before.
>
> He found a mouse was gotten in
> The hollow tube, and, shut between
> The two glass windows in restraint,
> Was swelled into an elephant,
> And prov'd the virtuous occasion
> Of all this learned dissertation:

Butler, the poet, then spoke his mind through the character of one of the more junior philosophers. "No wonder we're . . . the talk of the town . . . for all our great Attempts, we have done nothing yet." He criticised the scientists for dealing " . . . in novelty and change, / Not of things true, but rare and strange." Still, the elders preferred to see what they wanted rather than to see what was true.

> Some swore, upon a second view,
> That all they 'ad seen before was true,
> And that they never would recant
> One syllable of th' elephant;

At last the telescope was disassembled, and the inhabitants of the Moon were found to be gnats and flies entrapped in the tube. And the mouse scampered free when the lens was removed. Butler then described the real discovery:

> That those who greedily pursue
> Things wonderful instead of true,

> That in their speculations choose
> To make discoveries strange news,
> And natural history a Gazette
> Of tales stupendous and far-fet;
> Hold no truth worthy to be known
> That is not huge and overgrown;
> And explicate appearances
> Not as they are, but as they please;
> In vain strive nature to suborn,
> And for their pains, are paid with scorn.
>
> -from "The Elephant in the Moon"-
> -Samuel Butler-

John Keats (1745-1821) represented the Romantic poets who perferred myth and mystery to the harsher realities of truth that were being exposed by science. They were unforgiving of the scientist who bespoiled their idealistic existence. Keats wrote that "I have loved the principle of beauty in all things." It is no small wonder that mechanical truth was so difficult for him to accept.

> Do not all charms fly
> At the mere touch of cold philosophy (science)?
> There was an awful rainbow once in heaven:
> We know her woof, her texture; she is given
> In the dull catalog of common things.
> Philosophy will clip an Angel's wings,
> Conquer all mysteries by rule and line,
> Empty the haunted air and gnomed mine-
> Unweave a rainbow..."
>
> - from "Lamia II" -
> - John Keats -

Disapproval of the new science was not limited to the poet. George John Romanes was a zoologist during the time of Charles Darwin, but one who did not consider himself to be in the same camp as the evolutionists. He was convinced that truth was not the exclusive property and domain of science. He chose to look elsewhere for the entrance to that Citadel.

> I saw in dreams the Citadel of Truth -
>   A place as of polished silver, wrought
> With precious stones. Three god-like forms of youth
>   Attempted entrance. First a giant sought
> To force the door, beseiging it with blows.
>   He pausing, next a Child advanced with soft
> Inquest; and all about the pile he goes
>   In ceaseless gaze, around, adown, aloft.
> Last came a Greek-like maiden, fairy-bright,
>   Who held in both her hands a golden key;
> The lock was turned; in floods of rainbow light
>   I saw her pass; and then no more could see.
> Where Thought and Science access failed to win,
>   'Twas Art that opened, Art that entered in.
>   - "What Is Truth? "-
>   - George John Romanes -

Some writers consider it to be far from coincidental that the development and expansion of William James' (1842-1910) philosophy of pragmatism followed so closely on the heels of the subjective relativity of Einstein. If the truths of space and time were dependent upon the observer and his frame of reference, why would not other truths be dependent upon the observer and her frame of reference? The pragmatists believe that one cannot judge an idea simply by looking at it. An idea may be true under one set of circumstances, but false under another. The pragmatic philosophy, however, has been misinterpreted to mean that an idea was true if it enabled one to achieve what he wished. The founding proponents, William James and John Dewey, did nothing to encourage this interpretation. They insisted that truth was verified only when actions based upon it achieved the predicted results. A proposition was true only if it succeeded in linking the past with the present. Truth did not exist in isolation. In essence, pragmatism is the logic behind the scientific method.

Carl Sandburg presented the issue of truth in several of his poems. In each case he clearly recognized the existence of truth, but he also speculated as to whether it was within the capacity of man to recognize it. Can man judge truth simply by looking at it?

In *The People, Yes* a farmer is brought into the courtroom to testify in a particular case. As is the custom, he is sworn in by the attorney, and it is in the farmer's reply that pragmatism and relativity come face to face with truth.

> "Do you solemnly swear before the everlasting God
>   that the tesimony you are about to give in this
>   cause shall be the truth, the whole truth, and
>   nothing but the truth?"
>
> "No, I don't. I can tell you what I saw and what I
>   heard and I'll swear to that by the everliving
>   God, but the more I study about it the more sure
>   I am that nobody but the everliving God knows
>   the whole truth..."
>   - from *The People, Yes* #73 -
>   - Carl Sandburg -

In order to keep a perspective of the contradictions surrounding the quest for truth during the past two thousand years, we need only to refer again to the writings of the Roman, Lucretius. In addition to Divine Revelation, experimental science, relativity, and pragmatism, Lucretius stated most emphatically that "... truth originates out of the senses."

> .... You will find
> All knowledge of the truth originates
> Out of the senses, and the senses are
> Quite irrefutable. Find, if you can,
> A standard more acceptable than sense

> To sort out truth from falsehood. What can be
> More credible than sense? Shall reasoning,
> Born of some error, some delusionment,
> Argue the senses down? Ridiculous!
> If sense is false, reason will have to be.
> Can ears refute the eyes, the sense of touch
> Negate the sense of hearing? Do our noses
> Appeal against our eyes, our sense of taste
> File counterclaim against our ears' report?
> I'd hardly think so. To each sense belongs
> Its jurisdiction, so that soft, hot, cold,
> Color, sound, shape, and odor are assigned
> To different areas. Therefore, no sense
> Can contradict another or itself,
> Since their report must be dependable
> The same way always. If at any time
> A thing seems true to them, it must be so.
>
> - from *De Rerum Natura* IV, 483-499 -
> - Lucretius -

Robert Frost combined two ideas to describe the nature of truth. First he drew upon the words of Democritus, who wrote: "Of truth we know nothing, for it lies at the bottom of the well" -- difficult to see and difficult to reach. He combined that idea with the concept of relativity whereby the position of the observer was critical to seeing to the bottom of the well. This idea and the physics involved was examined previously in the chapter dealing with light.

In Frost's poem, "For Once, Then, Something," the observer chastises himself for always being in the wrong position to see the bottom of the well -- for never being able to see clearly the truth. His search inevitably resulted in only glare and reflections of himself and his surroundings. Once, only once, was he able to see through himself and catch a fleeting glimpse of truth, and even then there was uncertainty. Was it truth he saw, or was it just a pebble of quartz? Should truth be more dazzling than a pebble, or is truth "Truth" whatever it is?

> Others taunt me with having knelt at well-curbs
> Always wrong to the light, so never seeing
> Deeper down in the well than where the water
> Gives me back in a shining surface picture
> Me myself in the summer heaven godlike
> Looking out of a wreath of fern and cloud puffs.
> Once, when trying with chin against a well-curb
> I discerned, as I thought, beyond the picture,
> Through the picture, a something white, uncertain,
> Something more of the depths -- and then I lost it.
> Water came to rebuke the too clear water.
> One drop fell from a fern, and lo, a ripple
> Shook what ever it was lay there at bottom,
> Blurred it, blotted it out. What was that whiteness?
> Truth? A pebble of quartz? For once, then, something.
>   - "For Once, Then, Something" -
>   - Robert Frost -

Stephen Crane (1871-1900) agreed that truth is difficult to acquire, but he did not go so far as to say it was difficult to know. In one of his brief poems he considers truth from a very pragmatic view with the outcome being a cynical rationalization that truth is relative. If Truth represents danger, then there are, no doubt, less dangerous truths.

> The wayfarer,
> Perceiving the pathway to truth,
> Was struck with astonishment.
> It was thickly grown with weeds.
> "Ha," he said,
> "I see that none has passed here
> In a long time."
> Later he saw that each weed
> Was a singular knife.

> "Well," he mumbled at last,
> "Doubtless there are other roads."
> - "The Wayfarer" -
> - Stephen Crane -

Then what of truth? If truth is absolute, then why is it so illusive? If truth is changing, why is it so pursued? If truth is so important, why should one follow Emily Dickinson's admonition to "Tell all the Truth but tell it slant-"?

What then is the importance of truth in the relativistic and pragmatic world that science has thrust upon twentieth century man? And what of science itself? Is science the enterprise of hope for man or the messenger of his doom?

> Science, that gives man hope to live without lies
> Or blast himself off the earth:--curb science
> Until morality catches up?--But look: morality
> At present running rapidly retrograde,
> You'd have to turn science too, back to the witch-doctors
> And myth-drunkards. Besides that morality
> Is not an end in itself: truth is an end.
> To seek the truth is better than good works, better than survival,
> Holier than innocence and higher than love.
> - "Curb Science? " -
> - Robinson Jeffers -

CHAPTER X   ACCLAIM AND DISCLAIM

> From the nineteenth century view of science as
> a god, the twentieth century has begun to see it
> as a devil. It behooves us now to understand
> that science is neither one nor the other.
> - from *Education for a New Morality* -
> - Agnes Meyer -

The inescapable duality of science -- benefactor or malefactor of mankind -- has caused poets both to praise and condemn science. They have lavished science with almost god-like qualities -- this noblest of man's endeavors -- while simultaneously condemning its abandonment of diety, critcizing its degradation of man and nature, and decrying its irrelevance to man's true concens.

For the first sixteen centuries following the common era, science progressed slowly and somewhat unimpededly, albeit haltingly at times. The revelations of Copernicus, the denial of the geocentric universe, the trial of Galileo, the chemistry of Paracelsus, and the age of the earth brought science and religion into open conflicts, some of which have continued into the twentieth century. Amidst the conflicts we find the poet on both sides, and frequently at the same time. Acclaim and disclaim abound for this two-headed creature named science.

A major concern regarding science was its constant challenging of beliefs. Doubt and challenge became synonomous with science, and inevitably the doubts and challenges were directed toward religious beliefs. "To teach doubt and Experiment," wrote William Blake, "Certainly was not what Christ meant." Accordingly, Pope writes that man should rest within the comforts of the biblical explanations. Nothing of value could be gained by doubt.

>He who doubts from what he sees
>Will ne'er believe, do what you please,
>If the Sun and Moon should doubt,
>They'd immediately go out.
>   - from "Essay on Man" -
>   - Alexander Pope -

Biblical teachings and Reason were for centuries the source of truth. But that swing of the pendulum reached its maximum, and by the seventeenth century was reversing itself. Experimentation, which had been brought to the forefront by the essayist Francis Bacon in the early 1600s, became the wedge between the reasoning philosophers and the scientists as we know them today. No longer was philosophical reasoning the final word. Experimental science with its facts and logical development created and shattered philosophies and modified theology. And Isaac Newton led the way.

>You don't believe--I won't attempt to make ye:
>You are asleep--I won't attempt to wake ye.
>Sleep on, Sleep on! while in your pleasant dreams
>Of reason you may drink of Life's clear streams.
>Reason and Newton, they are quite two things:
>For so the Swallow & the Sparrow sings.
>Reason says "Miracle" ; Newton says "Doubt."

Aye! that's the way to make all Nature out.
"Doubt, Doubt, & don't believe without experiment"
        - from "Epigrams, Verses, and Fragments" -
        - Henry Jones -

As the number of experiments increased under the influence of Newton's new science, they too became the target of the poets' wrath. The data collections were scorned as being meaningless, and the conclusions were rebuked as absurdities. The misguided fools who produced them would someday recognize their folly.

Go, wond'rous creature! mount where Science guides,
Go, measure earth, weigh air, and state the tides;
Instruct the planets in what orbs to run,
Correct old Time, and regulate the Sun; . . .
Go, teach Eternal Wisdom how to rule--
Then drop into thyself, and be a fool!
        - from "Essay on Man" -
        - Alexander Pope -

Other poets, however, saw in the revelations of science a reinforced respect and admiration for the marvels of Creation and the wondrous omnipotence of God. Contrary to the idea that science was anti-god, they saw science as a buttress, if not proof, to the existence of God. Even the telescope evoked a spiritual reaffirmation of God.

My mightie wings high stretched then clapping light,
I brush the starres and make them shine bright.
Then all the works of God with close embrace
I dearly hug in my enlarged arms.
        - from "Cupid's Conflict" -
        - Henry More -

Thomas Chatterton (1752-1770) was another of the mid-eighteenth century poets for whom the revelations of science only intensified his respect for the divine creation. His poem, "The Copernican System," described the celestial wonders of the night sky. Rather than to condemn the orderliness and mechanical predictabilities of the heavenly bodies as being anti-God, he saw them as further evidence of the supreme creation. "These are thy wondrous works . . ." he wrote, "Now more admir'd in being understood."

> The Sun revolving on his axis turns,
> And with creative fire intensely burns;
> Impell'd the forcive air, our Earth supreme,
> Rolls with the planets round the solar gleam;
> First Mercury completes his transient year,
> Glowing, refulgent, with reflected glare;
> Bright Venus occupies a wider way,
> The early harbinger of night and day;
> More distant still our globe terraqueous turns,
> Nor chills intense, nor fiercely heated burns;
> Around her rolls the lunar orb of light,
> Trailing her silver glories through the night:
> On the Earth's orbit see the various signs,
> Mark where the Sun, our year completing, shines;
> First the bright Ram his languid ray improves;
> Next glaring wat'ry thro' the Bull he moves;
> The am'rous Twins admit his genial ray;
> Now burning, thro' the Crab he takes his way;
> The Lion, flaming, bears the solar power;
> The Virgin faints beneath the sultry shower.
>
> Now the just Balance weighs his equal force,
> The slimy Serpent swelters in his course;
> The sabled Archer clouds his languid face;
> The Goat, with tempests, urges on his race;
> Now in the water his faint beams appear,
> And the cold Fishes end the circling year.

Beyond our globe the sanguine Mars displays
A strong reflection of primeval rays;
Next belted Jupiter far distant gleams,
Scarcely enlight'ned with the solar beams;
With four unfix'd receptacles of light,
He tours majestic thro' immensity of space.
  These are thy wond'rous works, first Source of good!
Now more admir'd in being understood.
            - "The Copernican System" -
            - Thomas Chatterton -

These new "Engines," the telescope and microscope, were not accepted enthusiastically by all scholars and poets. Johann Goethe wrote, "Microscopes and telescopes...put our human eyes out of their natural, healthy and profitable point of view." Still, the expanded universe, as seen through the telescope, opened not only new visual vistas, but also spiritual ones. Perhaps man was not alone in the vastness. Such thoughts appeared to some poets as blasphemy to the words of Genesis, but to others they described an even more paternal diety.

Perhaps a thousand other worlds that lie
Remote from us, and latent in the sky,
Are lighted by his beams, and kindly nursed.
            - from "Elenora" -
            - John Dryden -

The passing centuries have exacerbated the dilemma of the duality of science. The American poet, Robinson Jeffers (1887-1962), acknowledges the achievements of science, saying, however, that these remarkable achievements and incredible creations may well be knives turned inward. The irony to Jeffers was that man's knowledge is so meagre when compared to all that is still unknown that it is as "a pebble on the shingle." "Who would have dreamed this infinitely little too much?" Can man's scientific genius survive his ignorance?

> Man, introverted man having crossed
> In passage and but a little with the nature of things this
>   latter century
> Has begot giants; but being taken up
> Like a maniac with self-love and inward conflicts cannot
>   manage his hybrids.
> Being used to deal with edgeless dreams,
> Now he's bred knives on nature turns them also inward:
>   they have thirsty points though.
> His mind forbodes his own destruction;
> Actaeon who saw the goddess naked among the leaves
>   and his hounds tore him.
> A little knowledge, a pebble from the shingle,
> A drop from the oceans: who would have dreamed this
>   infinitely little too much?
>   - "Science" -
>   - Robinson Jeffers -

The purpose and value of science

The duality of science seems inescapable -- blessing or curse? What is its value? What is to be gained from an unceasing search leading to uncertain findings. Albert Einstein wrote, "No amount of experimentation can ever prove me right; a single experiment may at any time prove me wrong." Is this uncertainty the epitomy of man's great efforts?

The question of the value of scientific knowledge has been asked for centuries. The diversity of responses has been great. The Arabian poet, Mu'tamid (1040-1095), gently counseled those who would search for truths, that knowledge is an "inconsistent thing" -- elusive at best. Even should one grasp the wisdom, it is as a "robe of dust" -- impermanent and fleeting.

> Woo not the world too rashly, for behold,
>   Beneath the painted silk and broidering,
>   It is a faithless and inconstant thing.
> (Listen to me, Mu'tamid, growing old.)
> And we--that dreamed youth's blade would never
>   rust,
>   Hoped wells from the mirage, roses from the sand--
>   The riddle of the world shall understand
> And put on wisdom with the robe of dust.
>   - "Woo Not the World" -
>   - Mu'tamid -

By the seventeenth century some poets had become annoyed and disturbed by what they considered to be the trivial investigations of scientists. John Donne issued a call for restraint. "We see in Authors too stiffe to recant,/ A hundred controversies of an Ant." He itemized the unsettled questions of the scientists and the issues that continued to escape their efforts. He suggested that these were insignificant in comparison to the real quest of man -- to know thyself.

>   Have not all soules thought
> For many ages, that our body'is wrought
> Of Ayre, and Fire, and other Elements?
> And now they thinke of new ingredients,
> And one soule thinkes one, and another way
> Another thinkes, and 'tis an even lay.
> Knowst thou but how the stone doth enter in
> The bladders Cave, and never breake the skin?
> Knowst thou how thy lungs have attracted it?
> There are no passages so that there is
> (For ought thou knowst) piercing of substances.
> And of those many'opinions which men raise
> Of Nailes and Haires, dost thou know which to praise?

> What hope have we to know our selves, when wee
> Know not the least things, which for our use bee?
> - from "Epicedes and Obsequies" -
> - John Donne -

The tone became more strident as the decades followed. More poets became concerned about where the scientific search for truth was leading. Clearly there was a point beyond which man should not venture, as Akenside wrote in his eighteenth century poem.

> There Science! Veil thy daring eye,
> Nor dive too deep, nor soar too high.
> - from "Hymn to Science" -
> - Mark Akenside -

Geofrey Whitney used the tale of Thales, the sixth century B. C. Greek philosopher, to reproach the scientists for their concern with irrelevancies while the consequential issues went unattended. In Diogenes' account, Thales stumbled into a ditch while observing the stars, and was chastised by an old woman. "How could he be expected to understand the heavens when he could not see under his feet?"

> Th' ASTRONOMER by night beheld the stars to shine:
> And what should chance an other year began for to divine.
> But while too long in skies the curious fool did dwell,
> As he was marching through the shade, he slipt into a well.
> Then crying out for help, had friends at hand, by chance;
> And now his peril being past, they thus at him do glance.

> What foolish art is this (quoth they) thou hold'st so dear,
> That thou foreshadow the perils far: but not the dangers near?
> - "Astronomer" -
> - Geofrey Whitney -

The barbed wit of Jonathan Swift (1667-1745) typified the impressions of many poets about the triviality of scientific investigations and the meaninglessness of their findings.

> So, naturalists observe, a flea
> Has smaller fleas that on him prey;
> And these have smaller still to bite 'em,
> And so proceed ad infinitum.
> - from "On Poetry" -
> - Jonathan Swift -

Johathan Swift was a man with little love for science. As the scholar Ifor Evans wrote, "Swift showed no comprehension whatever for the aim and purpose of science . . . He regarded scientific investigation as pretentious, merely a symptom of human vanity, and as such he brought all his masterly resources of venom into the attack, without making any effort to discover what it was all about."[1] In fairness to Swift, however, his attack on science was only a part of his attack on all forms of man's vanity and pretensions.

One of Swift's more sarcastic attacks on science, scientists, and their organizations is to be found in *Gulliver's Travels*. Gulliver was permitted to visit the grand Academy of Lagado, and to observe the "Arts wherein the Professors employ themselves." Afterwards he commented on the importance to society of such an institution.

> . . . there is not a Town of any Consequence in the

Kingdom without such an Academy (of science). In these Colleges, the Professors contrive new Rules and Methods of Agriculture and Building, and new Instruments and Tools for all Trades and Manufactures, ... The only inconvenience is that none of these Projects are yet brought to Perfection, and in the mean time, the whole Country lies miserably waste.[2]

Swift, through Gulliver, verbally jabbed at the Professors during the fictitious tour of the Academy. He referred to the Professors' (the scientists') incessant need for money to support their investigations, and "the Practice of Begging from all who go to see them."[3] His most pointed sarcasm was directed toward their projects which he cynically catalogued for the reader.

> His (the Professor's) Employment ... was an Operation to reduce human Excrement to its original Food, by separating the several Parts, removing the Tincture which it receives from the Gall, making Odour exhale, and scumming the saliva.
>
> There was a man born blind ... (whose) Employment was to mix Colours for Painters ... (distinguishing them) by feeling and smelling.
>
> Others softening Marble for Pillows ...
>
> ... to propagate a breed of naked sheep.
>
> ... extracting Sun-Beams out of cucumbers, which were to be put into Vials hermetically sealed, and let out to warm the Air in raw inclement Summers.[4]

To reiterate, no group escaped the sarcasm of Swift -- scientist, mathematician, politician, soldier, or woman. All were the targets of his wit and wrath.

Implicit within the rationale for science is the presumption that mankind will be improved, and his quality of life will be enhanced. Science is to be the benefactor of man. Yet the previously mentioned duality is inescapable. Thomas Campbell (1774-1844) was one for whom the goodness of science was not obvious. To him, the endless doubting and searching did not carry man to any heights of well-being or to any new levels of moral conduct, but only dragged him down into deepening pits of despair.

> Are these the pompous tidings ye proclaim,
> Lights of the world, and demi-gods of Fame?
> Is this your triumph--this your proud applause,
> Children of Truth, and champions of her cause?
> For this hath Science searched on weary wing
> By shore and sea each mute and living thing?
> Launched with Iberia's pilot from the steep,
> To worlds unknown, and isles beyond the deep?
> Or round the cope her living chariot driven,
> And wheeled in triumph through the signs of Heaven?
> Oh! star-eyed Science, hast thou wandered there,
> To waft us home the message of despair?
> - from "The Pleasures of Hope" -
> - Thomas Campbell -

Science has had an attraction to the common man. Apart from its contribution to his material well-being, science has had an entertaining quality and often an anticipated spiritual benefit. His fascination with science was a bitter pill for the purist who looked upon science as pagan, empty, and dehumanizing. William Wordsworth was one poet who not only cared little for science, but also could not believe that anyone could derive any measure of satisfaction from science and its machines. His poem, "The Star Gazers," pictures the throng who paid for the opportunity to peer through a telescope for what they expect to be a glorious and spiritually awakening experience -- " . . . What an insight must it be!" He then describes their disappointments; they "Seem to meet with little gain, seem less happy than before." Each is seen to leave slowly -- troubled, unfullfilled, and dissatisfied as if science is not the answer.

>What crowd is this? what have we here! we must not
>   pass it by;
>A telescope upon its frame, and pointed to the sky:
>Long is it as a Barber's Poll, or Mast of little Boat,
>Some little Pleasure-Skiff, that doth on Thames's
>   waters float.
>The Show-man chuses well his place, 'tis Leicester's
>   busy Square;
>And he's as happy in his night, for the heavens are
>   blue and fair;
>Calm, though impatient is the Crowd; Each is ready
>   with the fee,
>And envies him that's looking--what an insight must
>   it be!
>Yet, Show-man, where can lie the cause? Shall thy
>   Implement have blame,
>A Boaster, that when he is tried, fails, and is put to
>   shame?
>Or is it good as others are, and be their eyes in fault?
>Their eyes, or minds? or, finally, is this resplendent
>   Vault?
>Is nothing of that radiant pomp so good as we have here?
>Or gives a thing but small delight that never can be dear?
>The silver Moon with all her Vales, and Hills of mightiest
>   fame,
>Do they betray us when they're seen? and are they but
>   a name?
>Or is it rather that Conceit rapacious is and strong,
>And bounty never yields so much but it seems to do
>   her wrong?
>Or is it, that when human Souls a journey long have had,

And are returned into themselves, they cannot but be sad?
Or must we be constrain'd to think that these Spectators rude,
Poor in estate, of manners base, men of multitude,
Have souls which never yet have ris'n, and therefore prostrate lie?
No, no, this cannot be--Men thirst for power and majesty!
Does, then, a deep and earnest thought the blissful mind employ
Of him who gazes, or has gazed? a grave and steady joy,
That doth reject all shew of pride, admits no outward sign,
Because not of this noisy world, but silent and divine!
Whatever be the cause, 'tis sure that they who pry and pore
Seem to meet with little gain, seem less happy than before:
One after One they take their turns, nor have I one espied
That doth not slacky go away, as if dissatisfied.
   - "Star Gazer" -
   - William Wordsworth -

  The questioning of the purpose of science, like so many questions that have plagued the poets, has escaped an absolute answer. What is the true purpose of physical science and inquiry? Even at the moment of Newton's greatness, when his genius was being widely acclaimed, the question lurked beneath the surface, popping to the top like bubbles of swamp gas in a marshy pool.

Could he, whose rules the rapid Comet bind,
Describe or fix one moment of his mind?
Who saw its fires here rise, and there descend,
Explain his own beginning, or his end?
Alas, what wonder!
   - from "Essay on Man" II, 35-39 -
   - Alexander Pope -

The twentieth century has brought the poet no closer to the answer. W. H. Auden (1909-1973) expressed the concern not only of poets, but also of persons outside the circle of science and scientists.

He reflects upon the physicists' model of the universe. The continuous, chaotic, kinetic molecular motion; cosmic radiation; the emptiness of solid matter; and matter and anti-matter are symbolized when he speaks of a "lover's kiss" and "an indeterminant gruel / Which is partly somewhere else."

Auden also questions man's place in the universe. He prefers "a geocentric view," and the idea of a finite universe, but acknowledges them to be "Exploded myths." Still he is uncomfortable with the idea of the Big Bang and the Steady State theories of creation. He expresses this feeling in the lines: " . . . but who / Would be at home a-straddle / An ever expanding saddle?"

He, like so many of his predecessors, questions the purpose and value of scientific inquiry. Clearly, he could rejoice more if he knew what science was looking for. He is just not sure that man was meant to take unto himself every scientific question. He concludes by saying that whether man should or should not inquire is something that "we shall learn."

> If all a top physicist knows
> About the truth be true,
> Then, for all the so-and-sos,
> Futility, and grime
> Our common world contains,
> We have a better time
> Than the Greater Nebulae do
> Or the atoms in our brains.
>
> Marriage is rarely bliss
> But, surely, it would be worse
> As particles to pelt
> At thousands of miles per sec

Around a universe
In which a lover's kiss
Would either not be felt
Or break the loved one's neck.

Though the face at which I stare
While shaving it be cruel,
Since year after year it repels
An aging suitor, it has,
Thank God, sufficient mass
To be altogether there,
Not an indeterminate gruel
Which is partly somewhere else.

Our eyes prefer to suppose
That a habitable place
Has a geocentric view,
That architects enclose
A quiet, Euclidean space--
Exploded myths, but who
Would feel at home a-straddle
An ever-expanding saddle?

This passion of our kind
For the process of finding out
Is a fact one can hardly doubt,
But I would rejoice in it more
If I knew more clearly what
We wanted the knowledge for--
Felt certain still that the mind
Is free to know or not.

It has chosen once, it seems,
And whether our concern
For magnitude's extremes
Really becomes a creature

> Who comes in a median size,
> Or politicizing nature
> Be altogether wise,
> Is something we shall learn.
> > - "After Reading a Child's Guide
> >   to Modern Physics" -
> > - W. H. Auden -

Why must there be within science the restless search? Why cannot men live quietly and serenely with the limitations of a finite world and an infinite universe? God's work is perfect, and his Creation should be a comfortable abode for man. One should accept the blessing. These are the things George Herbert seems to be saying in his poem, "Content," written in the early seventeenth century.

> This soul doth span the world, and hang content
> > From either pole unto the center:
> Where in each room of the well furnished tent,
> > He lies warm, and without adventure.
> > > - from "Content" -
> > > - George Herbert -

But just as the fascination for the search for new knowledge has been the opiate for the men of science, so too has it addicted the spirits of some poets. Herbert, himself, became captivated by the discoveries and theories of "the new Philosophy." Later one finds him writing glowingly of the new "wayes of Learning." He was adamant, however, to emphasize that new knowledge in no way lessened his belief of, and reverence to, God as the Creator.

> > ... what the starres conspire,
> What willing nature speaks, what forc'd by fire;
> Both th' old discoveries, and the new-found seas,
> > The stock and surplus, cause and historie:

> All these stand open, or I have the keys:
> Yet I love thee.
> > - from "The Pearl" -
> > - George Herbert -

The twentieth century with its more pragmatic poets presents a different response to the duality of science. Science is a potential force -- no more, or no less. It enables man "to live without lies \ Or blast himself off the earth." Science is amoral; man is the determiner of its goodness or badness. Should science be halted until man learns how to treat himself and his fellow man? No. "Truth is better . . ., better than survival."

> Science, that gives man hope to live without lies
> Or blast himself off the earth:--curb science
> Until morality catches up? --But look : morality
> At present running rapidly retrograde,
> You'd have to turn science too, back to the witch-doctors
> And myth-drunkards. Besides that morality
> Is not an end in itself : truth is an end.
> To seek the truth is better than good works, better than survival,
> Holier than innocence and higher than love.
> > - "Curb Science?" -
> > - Robinson Jeffers -

Acclaim

In spite of their questioning the value of science, poets have been generous in their acclaim of scientists and their works. In the previous chapter we have read Archibald McLeish's captivating tribute in "Einstein." Earlier we read poems devoted to scientific principles of entropy, thermodynamics, and gravitation, to name a few. One would be remiss, however, if two fascinating tributes to Sir Isaac Newton were not examined. The first was written by James Thompson in 1727, the second by Jean Theophile Desaguliers in 1728. But before reading them, one must reconsider the developments of science at that time.

Planetary orbits had been described; Kepler's Laws were adopted; and Newton's laws of motion and his theory of gravitation were unchallenged. All of these principles developed the concept of mathematical balance. Newton's Theory of Gravitation described the attractive force operating between any two bodies, and his Third Law of Motion was a clear statement of balance -- for every action there is an opposite and equal reaction. His concept of attraction and balance dominated the day. It was within this context that the poets created their tributes to Newton.

Thompson's poem, "To the Memory of Sir Isaac Newton,", is almost encyclopedic in its references to his genius and his new science. The poem is rich in its explanations of the implications of science to man's knowledge and understanding of his physical universe. He begins by emphasizing that Newton did not create the relationships. His genius was that of discovering the secrets of God.

> Shall the great soul of Newton quit this earth
> To mingle with his stars . . . .
>
> Who, while on this dim spot where mortals toil
> Clouded in dust, from motionless simple laws
> Could trace the secret hand of Providence,
> Wide-working through this universal frame.
>
>   Nature herself
> Stood all subdued by him and open laid
> Her every latent glory to his view.

Thompson continued with his praise of Newton. First he painted the picture of the tireless investigtor and his uniqueness among men. From that, Thompson recounted the Theory of

Gravity and the explanation of tidal action.

> But, bidding his amazing mind attend,
> And with heroic patience years on years
> Deep-searching, saw at last the system drawn,
> And shine, of all his race, on him alone.
>
> All intellectual eye, our solar round
> First gazing through, he, by the blended powers
> Of gravitation and projection, saw
> The whole in silent harmony revolve.
>
> Her every motion clear-discerning, he
> Adjusted to the mutual main and taught
> Why now the mighty mass of water swells
> Restless, heaving on the broken rocks,
> And the full river turning - til again
> The tide revertive, unattracted, leaves
> A yellow waste of idle sand behind.

Thompson writes of Newton's experimental methodology, his ceaseless study of the heavens, and his use of the telescope -- the "astronomic tube." Newton is thereby able to explain such apparent dissimilarities as the orbits of the planets, the tides, and the "stone projected to the ground" as a single "harmonic system." Indeed the magnificence of Newton was that from such a simple set of causes -- attractive forces -- he was able to describe "a complete universe."

> Then, breaking hence, he took his ardent flight
> Through the blue infinite; and every star,
> Which the clear concave of a winter's night
> Pours on the eye, or astronomic tube,
> Far stretching, snatches from the dark abyss,
> Or such as further in successive skies
> To fancy shine alone, at his approach

Blazed into suns, the living centre each
Of an harmonious system -- all combined,
And ruled unerring by that single power
Which draws the stone projected to the ground.
   O unprofuse magnificence divine!
O wisdom truly perfect ! thus to call
From a few causes such a scheme of things,
Effects so various, beautiful, and great,
An universe complete! And O beloved

After having first shown the Newton of the mathematical, mechanical simplicity of the universe, Thompson takes the reader into the laboratory of Newton. There he describes his work with light and the prism, and the discovery of the spectrum. Thompson carefully crafts the description of the spectrum in such a way as to deemphasize the mechanical "Unweaving of a tapestry," as Keats had condescendingly phrased it. Instead he spoke of a "gorgeous train of parent colors," and carried forth the visual imagery as he related the spectrum to the sunsets and the rainbows. He described it all as an "Infinite mass of beauty, ever flushing, ever new." Did ever a poet envision anything so fair?

Even Light itself, which every thing displays,
Shone undiscovered, till his brighter mind
Untwisted all the shining robe of day;
And, from the whitening undistinguished blaze,
Collecting every ray into his kind,
To the charmed eye educed the gorgeous train
Of parent colours. First the flaming red
Sprung vivid forth; the tawny orange next;
And next delicious yellow; by whose side
Fell the kind beams of all-refreshing green.
Then the pure blue, that swells autumnal skies,
Ethereal played; and then, of sadder hue,
Emerged the deepened indigo, as when
The heavy-skirted evening droops with frost;

> While the last gleamings of refracted light
> Died in the fainting violet away.
> These, when the clouds distil the rosy shower,
> Shine out distinct adown the watery bow;
> While o'er our heads the dewy vision bends
> Delightful, melting on the fields beneath.
> Myriads of mingling dyes from these result,
> And myriads still remain--infinite source
> Of beauty, ever flushing, ever new.
>   Did ever a poet image aught so fair,

Thompson's personal tribute peaks in the lines in which he rhetorically asks if anything could be more magnificent when compared to Newton and his work?

> But hark! methinks I hear a warning voice,
> Solemn as when some awful change is come,
> Sound through the world--' 'Tis done!--the measure's full;
> And I resign my charge.'--Ye mouldering stones
> That build the towering pyramid, the proud
> Triumphal arch, the monument effaced
> By ruthless ruin, and whate'er supports
> The worshipped name of hoar antiquity--
> Down to the dust! What grandeur can ye boast
> While Newton lifts his column to the skies,

As one can recognize, Thompson's tribute focused primarily upon the contributions of Newton toward mankind's understanding of the way the universal system worked. It emphasized indirectly the marvelous simplicity of God's creation, and how Newton proclaimed the perfection by linking the interrelatedness of phenomena to such "a few causes." Thompson's poem was a tribute to Newton and his model for the physical universe. The poetic tribute written by Desaguliers was completely different.

The title of Desagulier's poem describes accurately the thesis: *The Newtonian System of the World the Best Model of Goverment*.[5] Whereas Thompson stressed Newton's impact upon scientific thought, Desaguliers translated the Newtonian theories into politics. He insisted that "the discoveries of the *Principia* should be applied to human thinking in all fields...Newton had discovered all truth, had reformed all thinking; he had revealed the ways of God to man."[6]

Desagulier began his tribute with what he defined as "A plain and simple Account of the Systems of the World." The poet's first lines reminded the reader of an earlier era of purity in government "ere Bribry began / To taint the Heart of undesigning man." He proceeded to track the evolution of scientific thought and to commend various natural philosophers and their contributions, not withstanding their errors and limits.

He was quick to praise Pythagorus for leading manking from the darkness of ignorance and chaos to the dawning of the true science of astronomy. Pythagorus brought acceptance to the proposition of the orderliness of the heavens, and from this Desaguliers drew his first parallel of science and goverment. Pythagorus' order and its mythical "Musik of his Spheres," he said, "did represent / The ancient Harmony of Government."

> In ancient times, ere Bribry began
> To taint the Heart of undesigning Man,
> Ere Justice yielded to be bought and sold,
> When senators were made by Choice, not Gold,
> Ere yet the Cunning were accounted Wise,
> And Kings began to see with other's Eyes;
> Pythagoras his Precepts did reherse,
> And taught the System of the Universe;
> Altho' these Observations then were few,
> Just were his Reasonings, his Conjextures true:
> Men's Minds he from their prepossesions won,
> Taught that Earth a double Course did not run,

Diurnal round it self, and Ann'al around the Sun,
That the bright Globe, from his Aethered Throne,
With Rays diffuse on the Planets shone,
And, whilst they all revolv'd, was fixed alone.
What made the Planets in such Order move,
He said, was Harmony and mutual Love.
The Musik of his Spheres did represent
The ancient Harmony of Government:

Desaguliers continues with his praise of the earlier contributors to astronomical science. He writes of Ptolemy and his discovery of the variations in the orbits of the planets. He speaks of Galileo and his "new invented eyes," the telescope, "Whose Course destroyed the Ptolemaick World."

He continues his chronological parade and speaks glowingly of Copernicus, but also speaks directly to the limitation of the astronomer's work.

What praises to Copernicus are due,
Who gave the Motion, the Places, true;
But What the Causes of those Motions were,
He thought himself unable to declare.

He refers again much later to this limitation as "the Errors of Copernicus."

Beginning about midway the poem, Desaguliers begins his tribute to Newton, who he says had done more to bring about an understanding of the "Caelestial Science .... Than all the Sages that have shone before."

One of his first allusions to the relationship of Newton's theory to politics is found in line 130. He refers to the "Attractive Force" at work on the planets that "...turns their Motion from its devious Course," and is quick to point out that the Law alters their directions but at the same time leaves them free. He elaborates by saying that the Law "Directs but not Destroys, their Liberty." By inference the reader translates the lines to mean that civil law should similarly direct the actions, but not destroy the liberties of the people.

The gravitational attractions are then used to show the interplay among the planets and the sun, and to relate them to political interrelationships: "The Primaries . . . / attend their Chiefs / but respected the Sun." In their freedom, planets journey far and become more and more remote from the center of power. Yet the attractive force, which he calls "mutual Love," slows the planets' speed; stops their escape; and "Recalls the Wanders, who slowly move / at first, but hasten as they feel his Love." His thesis is reiterated: "Attractions (love) govern all the World's Machines."

> But Newton the unparallel'd, whose Name
> No Time will wear out of the Book of Fame,
> Caelestial Science has promoted more,
> Than all the Sages that have shown before.
> Nature Compell'd, his piercing Mind, obeys,
> And deftly shews him all the secret Ways,
> 'Gainst Mathematics he has no Defense,
> And yields t' experimental Consequence:
> His tow'ring Genius, from its certain Cause,
> Ev'ry Appearance, a priori draws,
> And shews th' Almighty Architect's unaltered Laws.
> That Sol self--pois'd in Aether does reside,
> And thence exerts his Virtue far and wide;
> Like ministers attending e'very Glance,
> Six Worlds sweep round his Throne in Mystic Dance.
> He turns their Motion from its devious Course,
> And bends their Orbits by Attractive Force,
> His Pow'r coerc'd by Laws, still leaves them free,
> Directs but not Destroys, their Liberty;
> Tho' fast and slow, yet regular they move,
> (Projectile Force restrained by mutual Love,)
> And reigning thus with limited Command,
> He holds a lasting Scepter in his Hand.
> By his Example, in their endless Race,

> The Primaries lead their Satellite,
> Who guided, not enslaved, their Orbits reun,
> Attend their Chiefs, but still respected the Sun,
>
> Salutes him as they go, and his Dominion own.
> Comets, with swiftness, far at distance, fly
> To seek remoter Regions in the Sky;
> But tho' from Sol, with rapid haste, they rolled,
> They more slowly as they feel the Cold;
> Languid, forlorn, and dark their state thy Moon,
> Despairing when in the Aphelion.
> But Phoebus, soften'd by their Penitence,
> On them benignly sheds his Influence,
> Recalls the Wanderers, who slowly move
> At first, but hasten as they feel his Love;
> To him for Mercy bend, sue, and prevail;
> Their Atoms crowd to furnish out their Tail.
> By Newton's help, tis evidently seen
> Attraction governs all the Worlds Machines.

The final verses caution that rulers should not fall victim to the "errors of Copernicus" -- not recognizing the forces that draw bodies together. One should be acutely aware of the parallel that exists between the attracting forces affecting planetary harmony and the attracting forces that affect political harmony. Newton's model gives one force the name of gravitational force, and gives the other the name of mutual love. "Governing by Fear, instead of Love" is a "jarring Motion" that can only shake "their Master's Throne."

The perfect model for governing should be Newton's Philosophy -- gravitational attraction in the planetary world, and mutual love in the political world. Love is the great force that binds "all the British Hearts."

> By now my cautious Muse consider well
> How nice it is to draw the Parallel:

Nor dare the Actions of the crown'd Heads to scan:
(At least within the Memory of Man)
If th' Errors of Copernicus may be
Apply'd ought within this Century,
When e'er the want of understanding Laws,
In government, might of some wrong Measures cause,
His Bodies rightly plac'd still rolling on,
Will represent our fix'd Succession,
To which alone the united Britons owe,
All the sure Happiness they feel below.
Nor let the whims of the Cartesian Scheme,
In politics be taken for thy Theme,
Nor say that any Prince should e'er be meant,
By Phoebus, in his Vortex, indolent
Suff'ring each Globe a Vortex of his own,
Whose jarring Motion shook their Master's Throne,
Who governing by Fear, instead of Love,
Comets, from ours, to others Systems drove.
But boldly let thy perfect Model be,
NEWTON'S (the only true) Philosophy:
Now sing of Princes deeply vers'd in Laws,
And Truth will crown thee with just Applause;
Rouse up thy Spirits and exalt thy Voice
Loud as the Shouts, that Speak for the People's Joys;
When Majesty diffuse Rays imparts,
And kindles Zeal in all the British Hearts,
When all the Powers of the Throne we see
Exerted to Maintain Liberty:
When Ministers within the Orbits move,
Honour their King, and shew each other Love,
When all Distinctions cease, except it be
Who shall the most excell in Royalty:
Comets from far, now gladly wou'd return,
And, pardon'd, with more faithful Ardour burn.
Attraction now in all the Realm is seen,
To bless the Reign of George and Caroline.

Mechanical science

The nineteenth century showed evidence of an increasing hostility toward science by some poets. The hostility was to a large extent caused by three factors. One factor was the movement of science in the direction of describing the universe in what seemed to them as totally mechanical terms. "God forbid," exclaimed poet William Blake, "that Truth should be Confined to a Mathematical Diagram . . . God is not a Mathamatical Diagram."[7]

A second factor was closely related to the first. Technology was even a more mechanical outgrowth of a mechanical science, and the industrial society was the ugly, pus-filled boil that scarred the lovely and serene face of the landscape. William Wordsworth captured these poets' feelings when he bemoaned that science and its civilization had put man "out of tune" with nature and God. He yearned for a former time and a return to the beauty he had known.

> Great God! I'd rather be
> A Pagan suckled in a creed outworn;
> So might I, standing on this pleasant lea,
> Have glimpses that would make me less forlorn.
> - from "The World Is Too Much with Us" -
> - William Wordsworth -

A third factor was even more difficult for the poets to accept -- the indifference of science. Scientists seemed to be completely unconcerned with the implications science had for society. Science progressed uninterrupted; it was oblivious to the beauty and imagery it crushed beneath its wheels of progress. Each new revelation appeared, at worst, to push God farther out of the picture. At best it seemed that science simply ignored with benign oversight the Creator's role.

Keats wrote of the abhorent demystification of natural phenomena by this growing monster called science. His description of science as "cold philosophy" described the mechanical, impersonal and insensitive destroyer of beauty. In his view, the beauty and mystery of the rainbow was degraded by science to a conglomeration of corpuscles of light much as a tapestry is unwoven into a litter of colored strings.

> ... Do not all charms fly
> At the mere touch of cold philosophy?
> There was an awful rainbow once in heaven:
> We know her woof, her texture; she is given
> In the dull catalogue of common things.
> Philosophy will clip an Angel's wings,
> Conquer all mysteries by rule and line,
> Empty the haunted air, and gnomed mine--
> Unweave a rainbow, as it erewhile made
> The tender-person'd Lamia melt into a shade.
> - from "Lamia" -
> - John Keats -

A similiar theme was expressed by the American poet, Edgar Allen Poe (1808-1849). Science, with its cool indifference and impersonal mechanical probing, was the antithesis of beauty, and a twisting dagger in the heart of the poet.

> Science! true daughter of Old Time thou art!
> Who alterest all things with thy peering eyes.
> Why preyest thou thus upon the poet's heart,
> Vulture, whose wings are dull realities?
> How should he love thee? or how deem them wise,
> Who wouldst not leave him in his wandering
> To seek for treasures in the jewelled skies,
> Albeit he soared with an undaunted wing?
> Hast thou not dragged Diana from her car?
> And driven the Hamadryad from the wood
> To seek a shelter in some happier star?
> Hast thou not torn the Naiad from her flood,

The Elfin from the green grass, and from me
The summer dream beneath the tamarind tree?
> - "Sonnet: To Science" -
> - Edgar Allen Poe -

Science was mechanical, and the scientists were unimaginative. Having this perception, some poets found it difficult to see little more than a dismal and forlorn existence for those who chose a life of scientific inquiry. Walt Whitman (1819-1892) wrote of these persons and of the misguided people who appeared to be seduced by the prostituted new knowledge. One reads within his words the expression of sorrow he has for those who have abandoned the reality of nature.

When I heard the learn'd astronomer;
When the proofs, the figures, were ranged in columns
    before me;
When I was shown the charts and diagrams, to add,
    divide, and measure them,
When I, sitting heard the astronomer, where he lectured
    with much applause in the lecture-room,
How soon, unaccountable, I became tired and sick;
Till rising and gliding out, I wander'd off by myself,
In the mystical moist night-air, and from time to time,
Look'd up in perfect silence at the stars.
> - "When I Heard the Learn'd Astronomer" -
> - Walt Whitman -

A quick look to the twentieth century and a poem by Diane Wakoski shows that Whitman's feelings are still shared by certain contemporary poets. She says, too, that mechanical science leads us to believe that we understand our existence because we have given names to it -- and thus we mistakenly believe that we know ourselves. Mechanical science ignores the true beauties, and love.

I have observed the learned astronomer
telling me the mythology of the sun.
He touches me with solar coronas.
His hands are comets with elliptical orbits, the
excuses for discovering planet X.
Lake water shimmering in sunset light,
and I think of the whitewashed dome of discovery
hovering over the landscape
wondering what knowledge does for us
in this old and beautiful un-knowing world.
                                        Yes,
                                            I would
rather name things
than live with wonder
or religion.
What the astronomer does not understand about poetry
is the truth of disguise.
That there are many names for the same phenomenon.
Love being
the unnamed
the unnameable.
              - "For Whitman" -
              - Diane Wakoski -

    The twentieth century has experienced to a far greater degree than any of the previous the effects of science, and the reduction of beauty and existence to mechanical explanations. Consequently, the hostility and frustration of some of the contemporary poets has risen to alarming levels as exemplified in the poem by Vachel Lindsay. Life, she exhorts, has been reduced by science to a kaleidoscope of gears, springs, and pipes. We are left, however, to interpret for ourselves the meaning of her final verse. Is it a vitriolic expression of wrath, or is it a more subtle illustration of our acceptance of science's devaluation of life?

> "There's machinery in the butterfly·
> There's a mainspring to the bee·
> There's hydraulics to a daisy,
> And contraptions to a tree.
>
> "If we could see the birdie
> That makes the chirping sound
> With x-ray, scientific eyes,
> We could see the wheels go round."
>
> *And I hope all men*
> *Who think like this*
> *Will soon lie*
> *Underground.*
>   - "The Horrid Voice of Science" -
>   - Vachel Lindsay -

Mechanical science and man's creative ingenuity combined to produce what has become known as the industrial revolution which in turn gradually evolved into the modern day technological society. Poets, as well as the population, at first accepted the new ways. But as the new ways began to surplant the old, and as the new ways seemed to eliminate the beauty and tranquility of the past, poets began to step forward and to view with alarm the products of science. We have read earlier the lines by William Morris as he attempted to return to a purer age:

> Forget six countries overhung with smoke,
> Forget the snorting steam and piston stroke,
> Forget the spreading of the hideous town;
> Think rather of the pack-horse on the down,
> And dream of London, small, and white, and clean,
> The clear Thames bordered by its garden green.
>   - from "Prologue to the Earthly Paradise" -
>   - William Morris -

Yet, what Morris viewed with alarm and distaste, other poets, such as Hart Crane have embraced as the glory of the new forces.

> The nasal whine of power whips a new universe...
> Where spouting pillars spoor the evening sky,
> Under the looming stacks of the gigantic power house
> Stars prick the eyes with sharp ammoniac proverbs,
> New verities, new inklings in the velvet hummed
> Of dynamos, where hearing's leash is strummed...
> Power's script, -- wound, bobbin-bound, refined--
> Is stropped to the slap of belts on booming spools...[8]

Two songs of the road, written some hundred years apart, demonstrate the change brought about by science as technology has increasingly intruded upon society. In the first we see that nature, beauty, and boundless freedom permeate Whitman's (1819-1892) "Song of the Open Road."

> Afoot and light-hearted I take to the open road,
> Healthy, free, the world before me,
> The long brown path before me leading wherever I choose.
>
> I inhale great draughts of space,
> The east and the west are mine, and the north and the
>   south are mine.
>
> Allons! whoever you are come travel with me!
> Traveling with me you find what never tires.
>   - from "Song of the Open Road" -
>   - Walt Whitman -

One hundred years of progress resulting from undreamed of advances in scientific and technological achievements has extorted its price. To some twentieth century poets the price is too high -- urban crush, ugliness, and restriction.

Yield.
No Parking.
Unlawful to Pass.
Wait for Green Light.
Yield.
Stop.
Danger.
Narrow Bridge.
Merging Traffic Ahead.
Yield.

Squeeze.
Dead End.
Do Not Enter.
Enter at Own Risk.
Yield.
Yield.
Yield.
Yield.

- "Song of the Road" -
- Ronald Gross -

Unquestionably, science and technology has produced a disturbing ambivalence, bringing with it an unsettling love-hate relationship. Phyllis McGinley captures the essence of these ambivalent feelings toward the god Technology.

I cannot love the Brothers Wright.
Marconi wins my mixed devotion.
Had no one yet discovered Flight
Or set the air waves in commotion,
Life would, I think, have been as well.
That goes for A. G. Bell.

What I'm really thankful for, when I'm cleaning up after Lunch, Is the invention of waxed paper.

That Edison improved my lot,
I sometimes doubt; nor care a jitney
Whether the kettle steamed, or Watt,
Or if the gin invented Whitney.
Better the world, I often feel,
Had nobody contrived the wheel.

On the other hand, I'm awfully indebted
To whoever it was dreamed up the elastic band.

Yes, pausing grateful, now and then,
Upon my prim, domestic courses,
I offer praise to lesser men--
Fultons unsung, anonymous Morses--
Whose deft and innocent devices
Pleasure my house with sweets and spices.

I give you, for instance, the fellow
Who first had the idea for Scotch Tape.

I hail the man who thought of soap,
The chap responsible for zippers,
Sun lotion, the stamped envelope,
And screens, and wading pools for nippers,
Venetian blinds of various classes,
And bobby pins and tinted glasses.

DeForest never thought up anything
So useful as a bobby pin.

Those baubles are the ones that keep
Their places, and beget no trouble,
Incite no battles, stab no sleep,
Reduce no villages to rubble,
Being primarily designed
By men of unambitious mind.

You remember how Orville Wright said his flying machine
Was going to outlaw war?

Let them on Archimedes dote
Who like to hear the planet rattling.
I cannot cast a hearty vote
For Galileo or for Gatling,
Preferring, of the Freaks of science,
The pygmies rather than the giants--

(And from experience being wary of
Greek geniuses bearing gifts)--

Deciding, on reflection calm,
Mankind is better off with trifles:
With Band-Aid rather than the bomb,
With safety match than safety rifles.
Let the earth fall or the earth spin!
A brave new world might well begin
With no invention
Worth the mention
Save paper towels and aspirin.

Remind me to call the repairman
About my big, new, automatically defrosting refrigerator
   with the built-in electric eye.
          - "Reactionary Essay on Applied Science" -
          - Phyllis McGinley -

Physical science and man

    "What about us?" asks May Swenson in her poem "The Universe." Is man the ultimate creation of God with dominion over all? Or is man merely a product of random atomic collisions -- is man a speck on the earth, as earth is a speck within the solar system, as the solar system is a speck within the universe?

Unquestionably, science has had an impact upon man's feelings about himself, and his role within the universe. During the greater part of recorded history, earth was considered to be the center of the universe, and man was the center of the earth; God created earth for man. The poet, George Herbert, clearly expressed the popular belief when he wrote: "Man is everything and more." Man had little doubt about his place in the divine structure.

> For us windes do blow;
> The earth doth rest, heav'n move, and fountains flow.
> Nothing we see but means our good,
> As our delight, or as our treasure;
> The whole is, either our cupboard of food,
> Or cabinet of pleasure.
> - from "Content" -
> - George Herbert -

Then, like a barbarian hoard, science ruthlessly invaded the comfortable beliefs, and struck down the reassuring philosophies. The finite universe became the infinite universe. The center of the heavens became the sun, and earth became little more than its circling rock. Stephen Crane expressed this feeling of uncertainty and insignificance in his poem written about 1899.

> A man said to the universe:
> "Sir, I exist."
> "However," replied the universe,
> "The fact has not created in me
> A sense of obligation."
> - Stephen Crane -

Other poets, however, have credited science with bringing forth the realization of a greater importance of man. Shelley was one who believed that science would permit mankind to cast out hate and fear, and to destroy the evils that had enticed him from the perfection of Creation. Through man's mastery of science he would control the forces of nature and himself. Shelley's epic poem, "Prometheus Unbound," allegorizes man's conflicts with the forces of nature, and predicts the time when science will enable man to overcome his expulsion, and to return to his state of perfection.[9]

> Man hath weav'd out net, and this net throwne
> Upon the Heavens, and now they are his owne.
>
> The lightening is his slave; heaven's utmost deep
> Gives up her stars, and like a flock of sheep
> They pass before his eye, are numbered, and roll on!
> The tempest is his steed, he strides the air;
> And the abyss shouts from her depth laid bare,
> Heaven, hast thou secrets? Man unveils me; I have none.
>   - from "Prometheus Unbound" -
>   - Percy Bysshe Shelley -

Again we have witnessed the duality of science. The science that some poets believe reduces man in size and importance, is the same science that other poets see as lifting man up to new levels of importance. The contemporary poet, Robinson Jeffers, writing in his obtuse manner, poemed the latter case.

>   Humanity is
>   the start of the race; I say
> Humanity is the mould to break away from, the crust to
>   Break through, the coal to break into fire,
> The atom to be split.
>
>   The atom bounds-breaking,
> Nucleus to the sun, electrons to planets, with recognition
> Not praying, self-equaling, the whole to the whole, the
>   microcosm
> Not entering nor accepting entrance, more equally, more
>   utterly, more incredibly conjugate

> With the other extreme and greatness; passionately
> perceptive of identity . . . .
>> - from "Roan Stallion" -
>> - Robinson Jeffers -

"Jeffers," writes Radcliffe Squires, "places man precisely in the central position which science tells us he occupies in the universe, midway in mass between the atom and the largest star. Does this dwarf man or magnify him? . . . . to me, Jeffers offers a vision of human participation in magnificence."[10]

"What about us?" Scientists appear reluctant to ask a question for which they are incapable of developing a quantitive answer. It is left, therefore, to the poets to continue to ask, "What about us?", and to continue to struggle for the answer. Don Marquis, another of the contemporary poets, implies his answer through a dialog with a toad.

> i met a toad
> the other day by the name
> of warty bliggens
> he was sitting under
> a toadstool
> feeling contented
> he explained that when the cosmos
> was created
> that toadstool was especially
> planned for his personal
> shelter from sun and rain
> thought out and prepared
> for him
>
> do not tell me
> said warty bliggens
> that there is not a purpose
> in the universe
> the thought is blasphemy

a little more
conversation revealed
that warty bliggens
considers himself to be
the center of the said
universe
the earth exists
to grow toadstools for him
to sit under
the sun to give him light
by day and the moon
and wheeling constellations
to make beautiful
the night for the sake of
warty bliggens

to what act of yours
do you impute
this interest on the part
of the creator
of the universe
i asked him
why is it that you
are so greatly favored

ask rather
said warty bliggens
what the universe
has done to deserve me
if i were a
human being i would
not laugh
too complacently
at poor warty bliggens
for similar
absurdities

> have only too often
> lodged in the crinkles
> of the human cerebrum
>> - "Warty Bliggens the Toad" -
>> - Don Marquis -

"What about us?" Poet May Swenson gives an answer in her poem, "The Universe," as it appears on the following page. Clearly, she asserts in the first part, man is extraordinary, and with his uniqueness comes the determination to understand the universe -- to "unspin / the laws that spin it."

In the second half of her poem, Swenson develops a strong instinctive feeling that more exists than man and the universe. It is in this portion of the poem that one becomes aware of possibly the most profound difference between the poet and the scientist.

The scientist, reading the lines of the poem, asks "What do you mean exactly? State clearly your message so that there will be no misunderstanding -- no ambiguity of meaning." The scientist expects exactitude. But for the poet, on the other hand, it is the feeling that is paramount. Samuel Coleridge explains the poet's reasoning when he writes, "Poetry gives most pleasure when only generally and not perfectly understood." Although Swenson has little hesitancy in saying that a proper role of man is scientific study, she feels no obligation to answer precisely the other questions, especially "What about us?" We, as the reader, must decide for ourselves the subtle meanings and physical array of her words Swenson's answer to the question. But then, is it her answer or ours?

                What
                    is it about,
                    the universe,
                    the universe about us stretching out?
        We, within our brains,
                within it,
                        think
    we must unspin
    the laws that spin it.
                We think *why*
            because we think
            because.
            Because we think,
                we think
                the universe about us.

            But does it think,
                the universe?
                    Then what about?
                            About us?
                    If not,
    must there be cause
                in the universe?
        Must it have laws?
                    And what
                if the universe
                        is not about us?
                    Then what?
                    What
                        is it about?
                    And what
                            about *us?*
                -"The Universe"-
                -May Swenson-

Physical science and poetry

Physical science and poetry would seem to have little in common. Science is thought of as being cold, calculating, and insensitive. Poetry is warm, imaginative, and romantic. Certain critics have, from time to time, asserted that the modern scientific point of view is inherently antagonistic to poetic inspiration and that the era of science could well witness the extinction of poetry. Yet, we have become aware that the persistent evolution of scientific theories has been paralleled by persistent discussions and interpretations of its pronouncements by poets.[11] Are physical science and poetry -- the two marvelous expressions of the creative human spirit-- isolated from each other?

> The true poet and the true scientist are not estranged. They go forth into nature like two friends. Behold them strolling through the summer fields and woods. The younger of the two is much the more active and inquiring; he is ever and anon stepping aside to examine some object more minutely, plucking a flower, treasuring a shell, pursuing a bird, watching a butterfly; now he turns over a stone, peers into marshes, chips off a fragment of a rock, and everywhere seems intent on some special and particular knowledge of the things about him. The elder man has more an air of leisurely contemplation and enjoyment,--is less curious about special objects and features, and more desirous of putting himself in harmony with the spirit of the whole. But when his younger companion has any fresh and characteristic bit of information to impart to him, how attentively he listens, how sure and discriminating is his appreciation! The interests of the two in the universe are widely different, yet in no true sense are they hostile or mutually destructive.[12]

Physical sciences and poetry: The twain meet.

References

Preface

1 Snow, C.P., *The Two Cultures and a Second Look*, New York: New American Library (1973), p. 11.

2 Newman, John Henry Cardinal, *The Idea of a University*, Oxford: Claredon (1976), p. 684.

Chapter I

1 Bronowski, Jacob, *The Ascent of Man*, Boston: Little Brown (1973), p. 227.

2 Levy, Hyman and Helen Spalding, *Literature for an Age of Science*, London: Methuen and Company Ltd. (1952), p. 85.

3 Levy, p. 84.

4 Waggoner, Hyatt Howe, *The Heel of Elohim*, Norman: University of Oklahoma Press (1950), p. 10.

5 Richards, I. A., *Poetries and Science*, London: Routledge & Keagan Paul (1970), p. 33.

6 Richards, p. 31.

7 Denny, Margaret, and William H. Gilman, (ed.), *The American Writer in the European Tradition*, New York: Haskell House Publishers (1968), p. 164.

8 Moore, F.J., *A History of Chemistry*, 3rd, New York: McGraw-Hill (1939), p. 164.

9 Wood, H.G., *Thought Life and Time As Reflected in Science and Poetry*, Cambridge: Cambridge University Press (1957), p.5.

Chapter II

1 Grabo, Carl, *A Newton among Poets,* New York: Cooper Square Publishers (1968), p. 140.

2 Unger, Leonard, and William Van O'Conner, *Poems for Study*, New York: Holt,

Rinehart and Winston (1961), p. 120.

3 Yoder, R.A., "Toward the 'Titmouse Dimension': The Development of Emerson's Poetic Style", *PMLA*, 87(2) March 1972, pp. 255-270.

4 Denny, Margaret, and William H. Gilman, (ed.), *The American Writer in European Tradition*, New York: Haskell House Publishers (1968), p. 154.

5 Rothenberg, Jerome, and George Quasha, *America: A Prophesy*, New York: Vintage (1964), p. 443.

Chapter III

1 Wing-Tsit Chan, (transl.), *A Source Book in Chinese Philosophy*, Princeton: Princeton University Press (1963), p. 151.

2 Lovejoy, Authur O., *The Great Chain of Being*, New York: Harper and Row (1960), p. 207.

3 Doyle, John Robert, Jr., *The Poetry of Robert Frost*, New York: Hafner Publishing (1962), p. 210.

4 Anderson, Charles, "Nothing Gold Can Stay", *The Explicator*, 22 (8), April 1964, p. 63.

5 Rifkin, Jeremy, *Entropy*, New York: Viking Press (1980), p.1.

6 Wing-Tsit Chan (transl.) *The Way of Lao Tzu (tao-te ching)*, Indianapolis: Bobbs-Merrill (1963), p. 151.

7 T'an Ssu-t'ung, "Ether and Humanity", from Wing-Tsit Chan (transl.), *A Source Book in Chinese Philosophy*, Princeton: Princeton University Press (1963), p. 739.

8 Gillis, Everett A., "Hope for Eliot's Hollow Men?", *PMLA*, 73 (5), December 1960, pp. 635-637.

9 Ryan, Lawrence V. and Friedrich W. Strothmann,"Hope for T.S. Eliot's "Empty Men'", *PMLA*, 73 (4), September 1958, pp. 426-432.

Chapter IV

1 Bynum, W.F. (ed.), *Dictionary of the History of Science,* Princeton: Princeton University Press (1981), p. 59.

2 Jacobi, Jolande (ed.), Norbert Guterman (transl.), *Paracelsus: Selected Writings*, New York: Panthenon Books (1958), p. 143.

3 Jacobi, p. 146.

4 Jacobi, p. 61.

5 Jacobi, p. 61.

6 Scerri, E.R., "The Tao of Chemistry", *Journal of Chemical Education*, 63 (2), February 1986, pp. 106-107.

7 Feng, Gia-Fu (ed.) and English, Jane (transl.), *Tao Te Ching*, New York: Random House (1975), p. 67.

8 Watts, Alan W., *Psychotherapy: East and West*, New York: Random House (1975), p. 106.

9 Chan, Wing-Tsit (transl.), *The Way of Lao Tzu*, Indianapolis: Bobbs-Merrill (1963), p.28.

10 Sellen, Eric, "Stevens 'The Glass of Water'", *The Explicator*, 17 (4), Janurary 1959, p. 28.

Chapter V

1 Todd, Ruthven, *Tracks in the Snow*, Baltimore: Johns Hopkins University Press (1977), p. 6.

2 Jones, William Powell, *The Rhetoric of Science*, London: Routledge and Keegan Paul (1966), p. 100.

3 Todd, p. 5.

4 Wasserman, Earl P., *The Finer Tone*, Baltimore: Johns Hopkins University Press (1953), pp. 173-174.

5 Wurtman, Richard, and Julia Wurtman,"Carbohydrates and Depression", *Scientific American,* 89, January 1989, p. 68.

6 Unger, Leonard (ed.), *Seven Modern American Poets*, Minneapolis: University of Minnesota Press (1967), p. 20.

7 Hart, Jeffery, "Frost and Eliot", *Sewanee Review*, 84 (3) Summer (1976), pp.442-443

8 Crowder, Richard, *Carl Sandburg*, New York: Twayne Publishers, Inc., (1964). p. 125

9 Crowder, p. 118.

Chapter VI

1 Hutchins, Robert Maynard (ed.), *Great Books of the Western World*, (Vol. 16, *Ptolemy, Copernicus, Kepler*), Chicago: Encyclopedia Britanica (1952), p. 1030.

2 Hutchins, p. 1049.

3 Ringler, W.A., *The Poems of Sir Phillip Sidney*, London: Oxford University Press (1962), p. 354.

Chapter VII

1 Haber, Francis C., *The Age of the World*, Baltimore: Johns Hopkins University Press (1950), p. 17.

2 Haber, p. 20.

3 Nicholson, Marjorie Hope, *The Breaking of the Circle*, New York: Columbia University Press (1960), p. 108.

4 Albritton, Claude C., Jr. (ed.), *Philosophy of Geohistory*, Stroudsberg, PA: Dowden, Hutchinson and Ross, Inc. (1975), p. 11.

5 Weber, Robert L., *More Random Walks in Science*, London: The Institute of Physics (1982), p. 62.

6 Murray, John J., "Frost's 'Accidently on Purpose'", *The Explicator*, 36 (2), Winter 1978, p. 17.

7 Cramer, Frederick H., *Astrology in Roman Law and Politics*, Philadelphia: American Philosophical Society (1954), p. 278.

8 Cramer, p. 232.

9 Parr, Johnstone, *Tamburlaines Malady and Other Essays on Astrology in Elizabethan Drama,* University, Al: University of Alabama Press (1953), pp. 55-57.

10 Parr, pp. 39-48.

11 Bok, Bart J. and Lawrence E. Jerome, *Objections to Astrology*, Buffalo, NY: Prometheus Books (1975).

12 Wood, H.G., *Thought, Life and Time*, Cambridge: Cambridge University Press (1957), p. 123.

13 Nicholson, Marjorie Hope, *Newton Demands the Muse*, Hamden, CT: Archon Books (1963), p. 27.

14 Deutsch, Babette, *Poetry in our Time,* New York: Columbia University Press (1956), p. 387.

15 Ellsworth, Mason, "Auden's 'As I Walked Out One Evening '", *The Explicator,* 12 (6), June 1954, p. 43.

Chapter VIII

1 Chan, Wing Tsit, *A Source Book in Chinese Philosophy*, Princeton: Princeton University Press (1963), p. 122.

2 Prince, Derek J. De Solla, *Science Since Babylon*, New Haven: Yale University Press (1961), p. 15.

3 Nakayama, Shigeru and Nathan Sivin, *Chinese Science*, Cambridge, MA: The MIT Press (1973), p. 79.

4 Nicholson, Marjorie, *The Breaking of the Circle*, New York: Columbia University Press (1960), p. 124.

5 Louthan, Doniphan, *The Poetry of John Donne*, New York: Bookman Associates (1951), p. 65.

6 Louthan, p. 66.

7 Lee, Desmond, *Plato: Timeas and Critias*, Baltimore: Penguin Books (1971), p. 42.

8 Olson, Richard, *Science Deified and Science Defied*, Berkeley: University of California Press (1982), p. 257.

9 Kerr, Richard A., "No Longer Willful Gaia Becomes Respectable", *Science*, 240, April 22, 1988, pp. 393-395.

10 Nicholson, p. 114.

11 Hutchins, John Maynard (ed.), *Great Books of the Western World*, (Vol. 16, *Ptolemy, Copernicus, Kepler*), Chicago: Encyclopedia Britanica (1952), p. 1051.

12 Hutchins, John Maynard (ed.), *Great Books of the Western World*, (Vol. 28, *Harvey*), Chicago: Encyclopedia Britanica (1952), p. 285.

13 Hutchins, (28), p. 285.

14 Hutchins, (28), p. 286.

Chapter IX

1 Falk, Signi Lenea, *Archibald MacLeish*, New York: Twayne Publishers (1965), p. 42.

2 Falk, p. 42.

3 Falk, p. 141.

4 Drew, Elizabeth, *Directions in Modern Poetry*, New York: Gordian Press (1975), p. 5.

5 Klosky, Milton (ed.), *Speaking Pictures*, New York: Crown Publishers (1975), p.5.

6 Soet, Mary Ellen (ed.), *Concrete Poetry: A World View*, Blomington: Indiana University Press (1968), p. 7.

7 Soet, p. 248.

Chapter X

1 Evans, Ifor, *Literature and Science*, London: George Allen & Unwin, Ltd. (1969), p. 36.

2 Swift, Jonathan. *Gulliver's Travels*, New York: Oxford University Press (1977), p. 174.

3 Swift, p. 172.

4 Swift, p. 176.

5 Desaguliers, Jean Theophile, *The Newtonian System of the World the Best Model of Goverment: An Allegorical Poem*, London: Westminister (1728).

6 Nicholson, Marjorie Hope, *Newton Demands the Muse*, Hamden, CT: Anchon Books (1963), p. 136.

7 Keynes, Geoffrey (ed.), *The Poetry and Prose Of William Blake*, London: The Nonesuch Library (1961), p. 806, p. 819.

8 Drew, Elizabeth, *Modern Poetry*, New York: Gordian Press (1940), p. 58.

9 Grabo, Carl, *A Newton Among Poets*, New York: Cooper Square Publishers (1968), p. 195.

10 Squires, Radcliffe, *The Loyalties of Robinson Jeffers*, Ann Arbor: University of Michigan Press (1955), p. 186.

11 Stevenson, Lionel, *Darwin among the Poets*, New York: Russell and Russell (1963), p.7.

12 Burroughs, John, *Indoor Studies,* Boston: Houghton Mifflin Company (1917), p. 72.

Index

Addison, John 44
Addison, Joseph 127
Adventures of Ideas 143
Africanus, Julius 144
Akenside, Mark 8, 252
Alchemist 70
alchemy 76, 76
Allston, Washington 94
All's Well 160
America: A Rhapsody 37
American Writer in the European Tradition 33
Anaximenes 18
Anderson, Charles 52
Antin, David 90
Apollonius 144
Aristotle 18, 39, 80, 123, 218,
Arnold, Matthew 11
astrology 157
astronomer 150
As You Like It 146
atomic theory 18, 224
atoms 18, 31, 80, 156
attractive forces 37, 38
Auden, W.H. 168, 172, 258
Augustus 157
Ayre, William 21, 42

Babcock, Matbie 132
Bacon, Francis 15, 246
Ball, Hugo 232
Bastard, Thomas 196
benzene ring 67, 78
Big Bang 151, 258
Biogenesis and Abiogenesis 146
Bishop, Morris 60, 220
Blake, William 9, 13, 23, 114, 246, 271
Bly, Robert 133
Bohr, Neils 22
Botanic Garden 68
Brache, Tycho de 123
Brereton, Jane 41
Bronowski, Jacob 8

Brooke, Henry 8, 166
Browne, Sir Thomas 146
Buffon, Comte de 147
Bushnell, Frances Louisa 122
Butler, Samuel 234

carbon dioxide cycle 204
Carlyle, Thomas 96
Carl Sandburg 114-119, 241
Carnot, Sadi 47, 52
Cawthorn, James 105
Chain of Being 6, 45, 165, 191, 205
chemicals 79
Cheney, John Vance 87
Chinese Science 56, 85, 194
Chung-shu, Tung 195
Churchhill, Winston 65
Clausius, Rudolf 47
Coleridge, Samuel 94
conservation of energy 191, 220
conservation of matter 181, 183, 220
continental drift 2, 168
Cooke, Frederick 111
Copernicus, Nicholaus 6, 146, 163, 245, 267
Coper, William 9, 106, 147, 212
Crane, Stephen 241
creation 39, 44, 49, 52, 108, 143, 145, 150, 165, 192, 202, 213, 215, 247
Crowder, Richard 114
Crowley, Robert 219

Dadaist 231
Dalton, John 8, 18
Dampier, Sir William 103
Darwin, Charles 149, 238
Darwin, Erasmus 55, 68
Davies, John 172
Dehn, Paul 33
Democritus 18, 32, 80, 241
Denham, John 176
De Rerum Natura 153, 174, 177, 182, 205, 213, 214

Desaguliers, Jean 265
Dewey, John 234
diastrophism 168, 216
Dickinson, Emily 93, 110, 242
Dictionary of the History of Science 68
disasters 42
Divine Institute 145
DNA 77
Donne, John 82, 123, 126, 195, 205, 234, 251
Doppler, Christian 152
Doppler effect 152
Doyle, John 51
Drummond, William 7
Dryden, John 6, 43, 249

earth, animate 203
earth quake 167
earth science 164, 167
earth's age 143
echo 137
elements 18, 77, 81, 195, 251
electricity 55, 57
electrostatic attraction 27
Einstein, Albert 60, 150, 194, 219, 250
Eliot, T.S. 61
Emerson, Ralph Waldo 22, 27, 42, 58, 96
end of the world 52, 146
energy 65, 220
entropy 47, 49, 52
Entropy 52
Ephisians 46
Euripedes 182
Evans, Ifor 253
evolution 8
experimental science 6, 33, 194, 240, 246, 251

Fernbach-Florsheim, Carl 232
Finer Tone 114
Flanders, Michael 53, 57
Fletcher, Phineas 196, 216
Frost, Robert 51, 83, 94, 113, 155, 157, 227, 242

Gaia 203

Galileo 163, 245, 267
Gamov, George 151
Genesis 103, 214
geocentric universe 245, 258, 280
geologists 148, 164
Gilbert, William Schwenck 96
Gillis, Everett 61
Gilpin, Bernard 96
Graves, Robert Ranke 67
gravitation 8, 34, 262, 267, 269,
Garbo, Carl 20
greenhouse effect 204
Gross, Ronald 79, 277

Hammond, William 81
Harmonice Mundi 123, 126, 217
Hart, Jeffrey 114
Harvey, William 203, 217
Hawthorne, Nathaniel 39
Heel of Elohim 13
helium 94
Helmholtz, Herman 50
Herbert, George 126, 260, 280
Hill, Aaron 49
History of Chemistry 14
Hood, Thomas 68
Hoyle, Fred 151
Hsun, Tsu 194
Huai, Nan Tzu 40
Hubble, Edwin Powell 150
Hubble's Law 150
Hutton, James 147, 201
Huxley, T.H. 121
hydrogen 100

Idea of a University 2
imagination 13
industrial society 271
industrialization 9
Island of Man 196

James, William 239
Jandl, Ernest 37
Jeffers, Robinson 242, 249, 261, 281
Jefferson, Thomas 65
Jennings, Soame 45
Johnson, Ben 70
Johnson, Samuel 96

Jones, Henry   247
Jones, William Powell   104
Julius Ceasar   49, 160

Keats, John   13, 109, 239, 271, 272
Kekule, Friedrich   14, 67
Kepler, Johanne   6, 123, 126, 146
Kimball, Hannah   21
kinetic energy   47
kinetic molecular theory   21, 27, 155, 258
King, Martin Luther Jr.   65
Kingston, Jeremy   227
Kipling, Rudyard   1, 5
Koestler, Arthur   122

Lacantius   145
Lao Tse   181
Lao tzu   40, 56, 85
Lao-Tzu   40, 56, 85
lead metal   70, 91, 93
Levy, Hymen   10
light 8, 103, 111, 166, 229, 243
Lindsay, Vachel   275
Lippershey   146
Literature for an Age of Science   10, 11
Louthan, Doniphan   198
love-making   198
Lucretius   153, 155, 174, 176, 182, 205, 213, 214, 242
Luke   47
Lyly, John   160
Luther, Martin   145

MacDiarmid, Hugh   77
McLiesh, Archibald   221
McGinley Phyllis   277
McKuen, Rod   136
magnet   95
magnetism   36, 95
Marlowe, Christopher   7, 81
Marquis, Don   282
matter   18, 179, 220
Matthew   46
mechanical science   271, 273
Meyers, Agnes   245
microcosm   195, 199, 216, 217, 218
Milton, John   128, 186, 205
More, Henry   247
Morgan, Robert   167
Morris, William   9, 275
motion   224
music of the spheres   123, 130, 265
Mu'tamid   251

Newman, Cardinal John   2
New Morality   245
Newton among Poets   20
Newton, Issac   7, 34, 40, 103, 110, 166, 220, 246, 262-270
Newtonian System of the World   265-70
Nicolson, Marjorie   195
nuclear   33, 91

"Objections to Astrology"   163
oceans   175
Ockerse, Tom   211
Optics   104, 166
Orchestra   172
order   193, 229, 248, 263
Origin of the Specie   8
oxidation   190

Pallister, William   164, 228
Paracelsus   69, 76
Paracelsus: Selected Writings   69
Paradise Lost   186, 205
Parr, Johnstone   158
Pearson, Norman   13, 22, 32
People, Yes   114-119, 241
phlogiston   67, 190
planetary motion   124, 186, 192, 268
planets   161
Plato   39, 81, 144, 202
Platonic Year   146
Poe, Edgar Allen   272
poems
  Accidentally on Purpose   156
  Action of Electricity   55
  After Great Pain   94
  After Reading a Child's Guide to Modern Physics   258
  Alchemist   70

All's Well   160
An Autograph   137
Arcades   128
Artillerie   126
As I Walked Out One Evening   186
As Soone as wee to bee   186
As You Like It   146
Astronomer   252
Atom from Atom   22
Auguris of Innocence   27, 14
Ballad of the East and West   1, 5
Bevard Fault   167
Cardinal Ideograms   153
Cloud/Horizon/Water   210
Cloud, The   207
Complaint: Or Night Thoughts on Life, Death and Immortality   89
Conflagration: An Ode   50
Content   260, 280
Copernican System   249
Cupid's Conflict   247
Curb Science?   249, 261
Definition for Mendy   91
De Microcosmo   196
De Rerum Natura   153, 174, 177, 182, 205, 213, 214
Desert Places   227
Design   44
Distances   227
$E=MC^2$ 60, 220
Earth   32
Earth and Sky   182
Einstein   221
Elenora   7, 249
Elephant in the Moon   235
Elegie XX   198, 199
Epicedes and Obsequies   253
Epigrams, Verses and Fragments   247
Epitaph Intended for Sir Issac Newton   103
Essay on Man   7, 41, 43, 45, 192, 234, 246, 247, 257
Ether and Humanity   57
Excursion, The   165
Fable of the Magnet and the Churn   97
Fire and Ice   51
First and Second law   53, 57

First Anniversary   82, 83, 232
For Once, Then, Something   113, 240
For Whitman   274
IV th Ecloques   125
Genesis   140, 214
Geology   164
Glass of Water   86
God by Their Names   162
Gold   68
Great Scarf of Birds   36
Happiest Heart   87
Hollow Men   61, 66
Holy Sonnets   195
Horrid Voice of Science   275
How Everything Happens   60
Hymn   44, 127
Hymn to Science   252
Ice Cream Cone   79
Innate Helium   94
In Time Like Air   98
Island of Man   196
It's Raining   178

Judgment Day   49
Karamane   223
Kind of Poetry I Want,   77
Lao-tzu   40
Lamia   109, 239, 272
Leg in the Subway   226
Light Exists in Spring   111
Like Attracts Like   37, 233
Love's Progress   198
Merchant of Venice   88, 125
Naked World   79
Newtonian System of the World   265-270
Night Thoughts of a Tortoise   220
Nothing Gold Can Stay   52, 84
Odes   137
On Poetry   253
On the Death of My Dear Brother   81
On the Late S.T. Coleridge   95
Orchestra   172
Order: A Poem   297
Oxygen   99
Paradise Lost   186, 205

Pearl, The   260
People, Yes   114-119, 241
Pleasures of Hope   253
Poems on Several Occasions   41
Progress of Love   176
Prolog to the Earthly Paradise
    9, 235
Prometheus Unbound   19, 130, 281
Purple Island   196
Queen mab   191
Reactionary Essay on Applied Science
    277
Regulations of Passions   105
Relativity   228
Rhymes for a Modern Nursery   33
Roan Stallion   281
Saturday August 27   233
Science   250
Seismograph   168
Self Reliance   58
Shadow of Judgement   7
Sickness   167
Silence Is Golden   136
Song of Myself   190
Song of the Open Road   276
Song of the Road   277
Sonnet: To Science   273
Soul and Sense   21
Sound of Silence   134
Spacious Firmament on High   127
Sphinx   31
Star Gazer   256
Tamburlaine   82
Tao-te ching   56, 85, 179
Task, The   106, 108, 147, 212
Tell All the Truth   244
Tempest, The   159
Ten Definitions of Poetry   17
There's a Certain Light   111
This Is My Father's World   132
3 Models of the Universe   153
To the Memory of Sir Isaac Newton
    262-265
Troilus and Cressida   159
Truth: Counterpart to Mr. Pope's
    Essay   21, 42
Twelfth Night   125
Two Cultures   1, 2
Tyger   230

Universe, The   279, 284
Universal Beauty   166
Upon the Kings Coronation   195
Warty Bliggens the Toad   282
Watching Television   133
Wayfarer   244
What Is Truth?   240
When Hydrogen Played Oxygen   100
When I Heard the Learn'd
    Astronomer   273
Woman in the Moon   161
Woo Not the World   251
World Is Too Much With Us,   271
World Music   120
Xenophanes   43
Poetries and Science   12
poetry,
    and science   11, 284, 286
    hostility of   9
    power of   11
Pope, Alexander   7, 9, 41, 43, 104,
    192, 215, 235, 246, 247,
    257
pragmatism   239, 242
Principia   7
prism   116
Psychotherapy: East and West   85
Ptolemy   6, 267
Purple Island   196
Pythagorus   123

rainbow   13, 238, 272
reflection of light   113, 241
refraction   166
relativity   60, 219, 224, 228, 233
Revolutions of the Celestial Spheres
    146
Rhetoric of Science   104
Richard III   160
Richards, I. A.   12, 21
Rieu, E. V.   220
Rifkin, Jeremy   52
rivers   173, 175, 176, 196, 216,
    217
Romanes, George John   240
Rowe, Elizabeth   49
Ruskin, John   165
rust   87
Ryan, Lawrence   61

salt 98
Sandburg, Carl 17, 23, 114, 241
Santayana, George 136
Sarton, May 98
Scerri, E. R. 84
science 230
    duality 33, 245, 249, 254, 255, 261, 277
    morality 243, 244
    poetry 284, 286
    politics 265, 267
    purpose 257, 284
    satire 33, 233, 254
    societies 252
    truth 239
senses 240
Seven Modern American Poets 113
Shakespeare, William 7, 81, 88, 97, 125, 146, 158
Shelley, Percy Bysshe 19, 130, 190, 207, 281
Sidney, Sir Phillip 125, 126
Simon, Paul 134
Smart, Christopher 9
Snow, C.P. 1
Soet, Mary Ellen 233
Solar System Theory of Atom 22, 281
sound waves 139
Source Book in Chinese Philosophy 57
space exploration 34
spectrum 114, 264
Spencer, Theodore 48
St. Luke 47
St. Matthew 46
Stanley, Thomas 95
Steady State Theory 151, 258
Stevens, Wallace 86
Strothman, Friedrich 61
Sullivan, A. M. 140, 168
Sully-Prudhomme 61
surrealism 233
Swedenborg, Emanual 27
Swenson, May 58, 60, 153, 284
Swift, Joan 99
Swift, Jonathan 253

Tamburlaine 82

Tamburlaine's Malady 158
T'an Ssu-T'ung 57
Tao-te ching 56, 85, 181
telescope 149, 234, 247, 249, 255, 263, 267
television 133
Tempest, The 159
Thales of Miletus 18, 252
Theophilus 144
thermodynamics 53, 57, 147
Thompson, Francis 35
Thompson, James 8, 261
Thompson, Lawrence 113
Thompson, Robert 156
Thompson, William 166
Thoreau, Henry David 139
Thought, Life and Time 15
tidal action 172, 263
Timaeus 202
Tracks in the Snow 106
Trolius and Cressida 159
truth 225, 241, 242, 243, 245
Twelfth Night 125

Updike, John 35
uniformitarianism 147, 214
universe 150, 153, 156, 202, 222, 224, 229, 245
Ussher, Archbishop James 143

Von Braun, Wherner 34

Waggoner, Hyatt Howe 11
Wakoski, Diane 274
Walden 139
water 87, 94, 98, 100
water cycle 176, 178, 183, 185, 196, 205, 207
Wasserman, Earl R. 110
Watts, W.A. 85
wave theory 57, 59, 133, 139
Way of Lao Tzu 57
Weaver, Mike 233
Webb, Francis 40
Wheelock, John Hall 32
Whitehead, Alfred North 143
Whitman, Walt 190, 273, 276
Whitney, Geofrey 252
Whittier, John Greenleaf 137

Williams, Emmett 37, 233
Williams, Oscar 226
Wither, George 162, 185
Woman in the Moon 161
Wood, H.G. 15
Wordsworth, William 13, 164, 256, 271
World War I 231

Yoder, R.A. 32
Young, Edward 89